botanic /bəˈtænɪk/

noun

- of or relating to botany or plants.
- designating or relating to herbal or botanical medicine.

Oxford English Dictionary, 3rd Edition

 IN THIS ISSUE

4	EDITORIAL KAREN M'CLOSKEY	68	FORENSIC ECOLOGIES AND THE BOTANICAL CITY MATTHEW GANDY
6	THE CHANGING NATURE OF BOTANIC GARDENS KATJA GRÖTZNER NEVES	74	BOTANIC LESSONS FROM THE PRAIRIE BERONDA L. MONTGOMERY
12	SPIRALING DIVERSITY AND BLANK SPOTS IN A 19TH CENTURY UTOPIAN BOTANIC GARDEN SONJA DÜMPELMANN	78	IN CONVERSATION WITH JARED FARMER KAREN M'CLOSKEY
22	GARDEN OF RELATION: DRAWING THE CLIMATIC INTELLIGENCE OF PLANTS BONNIE-KATE WALKER	86	IN CONVERSATION WITH PATRIC BLANC + CATHERINE MOSBACH KAREN M'CLOSKEY
32	PLANTS ON THE MOVE JANET MARINELLI	94	CONCEITS AND CONSTRUCTS: VEGETAL ARCHITECTURE ANNETTE FIERRO
38	GREEN GOLD: THE AKKOUB'S SETTLER ECOLOGIES IRUS BRAVERMAN	104	PLANT SAMPLES KAREN M'CLOSKEY
44	THE VAULT IS A BUNKER, THE ARSENAL ARE SEEDS XAN SARAH CHACKO	114	SMART PLANTS AND THE CHALLENGES OF MULTISPECIES NARRATIVE URSULA K. HEISE
52	DESIGN(ED) DECAY ANDREA LING		IMAGE CREDITS UPCOMING ISSUES
60	IN CONVERSATION WITH GIOVANNI ALOI KAREN M'CLOSKEY		

LA+ BOTANIC
EDITORIAL

From botanic classification to CRISPR technology, and from plant-robot sensors to debates about plant sentience, humans' relationship with plants continues to evolve. Plants permeate all aspects of our lives, whether used for food or medicine, extracted for scents or dyes, or placed in gardens or flowerpots. While interest in plants is nothing new for the professionals and hobbyists involved in gardening, horticulture, and botany—nor is it new to landscape architects—the following essays show there is much to examine regarding the assumptions that guide practice, particularly at a time of rapid climate change and species loss.

LA+ BOTANIC begins with botanic gardens, a typology dear to many landscape architects. Without eliding botany's ties to colonialism within which botanic gardens were emergent, anthropologist Katja Grötzner Neves explores how botanic gardens have become crucial to achieving just and sustainable futures through environmental education and food security. Looking back to the 19th century, historian Sonja Dümpelmann considers a botanic garden design by John Claudius Loudon. Using a spiral form with unfilled areas that could be extended *ad infinitum* to incorporate newly discovered plants, the garden was expressive of a deep-rooted belief in human progress. Dümpelmann asks how the spiral and "blank spots" remain relevant today as we acknowledge both rapid extinction and the impossibility of knowing all species. Continuing in the realm of design, landscape architect Bonnie-Kate Walker critiques the limited plant choices imposed by the horticultural industry, which not only impacts design but results in a lack of genetic diversity. Walker argues that these constraints are exacerbated by the plant hardiness zones delineated by the US Department of Agriculture and recommends augmenting them to include more climate criteria, particularly at a time when we must consider links between places and their future climates.

Reflecting on plant conservation efforts and the challenges posed by global heating and extinction, journalist Janet Marinelli considers assisted migration, which is controversial because it rebuffs definitions that have long been at the root of Western approaches to conservation. Marinelli argues that we have long been engaged in assisted migration by default—through the nursery trade of non-native ornamental plants. The two essays that follow also address problems with specific conservation approaches. Law professor Irus Braverman describes the intertwined natures of colonialism and capitalism, where environmental protections enforced by the state dispossess local and Indigenous communities. Historian of science Xan Sarah Chacko similarly argues how the methods and terminology of conservation can perpetuate past colonial practices, using seed banks as examples that serve as both reminders of human exploitation and as "collective" hope for human survival.

In contrast to seed banks, which preserve material in a suspended state, artist/architect Andrea Ling asks how we might learn to embrace the "leaky, smelly, and messy" by partnering with materials and embracing their agency and mutability as desirable qualities for how to pursue building our world. This question of material agency is also taken up in our interview with art historian/curator Giovanni Aloi. Aloi offers insights into an expansive range of contemporary art practices' engagements with plants and discusses the importance of art for stimulating questions about human-plant relations in ways that are less constrained than those of other disciplines.

The next grouping of essays and interviews asks what we can learn *from* plants, rather than what we can learn *about* them, especially those that thrive within human-dominated environments. Geographer Matthew Gandy calls this "forensic ecology" and looks at spontaneous urban flora in marginal sites. Plant scientist Beronda Montgomery calls this "lessons from plants" and argues that plants serve as inspiration for how we should cultivate human thriving. Environmental historian Jared Farmer sat down with *LA+* to discuss humans' relationships with elderflora, the oldest living trees that have survived millennia. Farmer reflects on how we might think about the elderflora of the future in the face of climate change. The third and final interview is with botanist Patrick Blanc and landscape architect Catherine Mosbach, who spoke with *LA+* about the differences between designing vertical gardens and designing public landscapes, and how they each try to cultivate an approach to plant life that is precisely structured yet open to spontaneity.

The last three essays focus on plant imagery in visual and textual representations. Architect Annette Fierro describes the changing nature of botanic imagery as represented in both drawn and built works of architecture, from its symbolic representation in ornamental patterning to buildings shrouded in living plants. My essay argues that recent technologies have led to the emergence of three types of "plant samples" in landscape imagery and that, despite being virtual, are often material in their effects. The issue ends with an essay that looks at possible botanic futures. Literary scholar Ursula Heise surveys the changing nature of plants in science fiction—from hostile adversaries to human-plant hybrids to fully sentient, central characters—a shift that illuminates our perceptions of the plant world and our place within it.

Karen M'Closkey
Editor in Chief

KATJA GRÖTZNER NEVES
THE CHANGING NATURE OF BOTANIC GARDENS

Dr. Katja Grötzner Neves is a professor of sociology and anthropology at Concordia University, Montreal, Canada. She is also axis director (with Dr. Jean-François Bissonnette) at the Quebec Center for Biodiversity Science. Neves has published extensively on the contemporary role of botanic gardens in governing biodiversity conservation. She has recently begun a new textbook on the sociology of environmental issues, centered on anti-racist decolonial pedagogy, as well as technology enhanced, student centered methods of engagement.

+ ANTHROPOLOGY

As we enter the third decade of the new millennium, the world faces the intensifying acuity of long-standing global challenges. Of direct relevance to this essay are matters of biodiversity loss, food security, and climate change. While much has been done in recent decades to tackle these challenges, the magnitude of their complexity has rendered substantive progress difficult. This, on the other hand, amplifies problems pertaining to social equity, political unrest, and geopolitical instability. The onset of the 2020 COVID-19 pandemic exacerbated these difficulties by introducing significant supply chain disruption, and by seriously disrupting economies on multiple levels.

It could be argued that these transformations offer opportunity to rebuild our socio-economic systems toward greater long-term resiliency. In fact, examples abound of projects developed precisely from this standpoint. More are being envisioned every day. Nevertheless, the immediacy of pressing concerns with macroeconomic stability—including high levels of inflation—has largely resulted in the political prioritization of short-term economic management measures. Prioritizing short-term approaches can be suboptimal in relation to global problems like biodiversity loss, food security, and climate change. In view of this state of affairs, it is more important than ever to consider the contributions that institutions other than official organs of state governance can make toward the mitigation of these global challenges in the medium to long term. Since the latter part of the 20th century, para-governmental, non-governmental, and private institutions have played increasingly consequential roles in addressing these issues. Among them, botanic gardens have played particularly significant leading roles.[1] Indeed, the range of botanical garden contributions to social and environmental issues is vast. And with close to 2,000 officially recognized botanic gardens spread around the globe, their reach and heterogeneity are quite considerable.

Botanic gardens today form a global network that strives to maintain and/or increase the planet's biological diversity, increase equitable access to food, as well as educate general audiences on environmental issues. While continuing to

play key roles as institutions of scientific research and public education, as well as remaining guardians of a rich variety of plant material, botanic gardens have also been instrumental in the development of global and national policy strategies for plant biodiversity conservation. In addition to this, many botanic gardens around the world have developed community engagement programs that help provide citizens with the means to produce edible plant foods and, in so doing, not only help increase food security but also contribute to the greater environmental sustainability of horticultural practices.

Though many botanic gardens still grapple with the conundrums that ensue from their embroilment with histories of colonialism and associated reduction of biological diversity around the world, dealing with these problems can constitute fruitful grounds for nurturing the kinds of holistic knowledge practices that will be critical to climate change mitigation and adaptation. One thing we have learned is that we will need deep systemic change on all fronts to secure the successful reordering of our societies toward sustainable futures. Such a shift is best secured by expanding our knowledge systems toward more holistic forms of thinking. The latter are better equipped to understand and respond to the high levels of complexity that characterize climate change phenomena.

The Global Preservation of Plant Biodiversity

The history of botanic gardens spans a period of over 500 years.[2] The rationale behind the creation of the modern botanical garden institution is twofold. On the one hand, they developed in reaction to the "discovery" of entire plant worlds that were unknown to Europeans prior to their arrival on the American continent at the end of the 15th century. On the other hand, botanic gardens owe their existence to the emergence of the scientific era, and the related commitment to know the world through objective systematic study. The inaugural mission of botanic gardens was to incorporate "newly discovered" plants into European plant classificatory schemes, which were built on Ancient Rome and Ancient Greek plant classification criteria. At this early state the focus was as much theological as it was scientific. The plant classification orders known to Europeans up until the 15th century were believed to reflect the full order of "God's earthly creations." The discovery of entirely new kingdoms of people, animals, and plants in 1492 that were new to Europeans required profound rethinking of these assumptions.

Soon after their inception in the 16th century, botanic gardens quickly established themselves as quintessential institutions of research and education, with plant systematics at the core of their endeavors. At this time the majority of botanic gardens that appeared in Europe were associated with universities. They were initially dedicated to the study and teaching of plant medicinal uses as the anchor point for training apothecaries and medical doctors. By the 17th century, however, botanic gardens had become full-fledged institutions for the study of plants in and of themselves under the rubric of modern scientific botany. Throughout the 17th, 18th, and 19th centuries, botanic gardens continued to expand their credentials as centers of plant knowledge production and dissemination. Under the guise of economic botany—which investigated the economic potential of plants—a large number of botanic gardens in the Global North became instrumental in the establishment of European colonialism and empire building projects. The independence of former European colonies from the 18th century onward—but especially throughout the 20th century—gave rise to a new wave of botanical garden expansion across the world in the service of newly created nation states.

Botanic Gardens Conservation International (BGCI), founded in 1987, recognizes the existence of 1,775 botanic gardens located in a total of 148 countries.[3] This list reflects new criteria BGCI implemented in 2018 concerning the definition of what constitutes a botanic garden, whereby greater emphasis is now placed in "conserving rare and threatened plants, compliance with international policies, and sustainability and ethical initiatives."[4]

In effect, a growing number of botanic gardens around the world became increasingly involved with attending to plant biological diversity and plant conservation throughout the

20th century and first two decades of the 21st century.[5] Some were instituted specifically to tackle biodiversity loss and/or to protect the endemic plants and ecosystems of the geographic areas in which they are situated. The *Jardim Botânico do Faial* in the Azores, Portugal, is one such example.[6] In keeping within its walls *Macaronesian* plants that are endemic and native to the Azores archipelago, the garden practices *ex situ* conservation. It also practices *in situ* conservation by engaging in ecosystem restoration initiatives on the island of Faial, as well as on the other eight islands that comprise the Portuguese archipelago of the Azores.

Other botanic gardens transformed themselves from former supporters of colonial practice to contemporary institutions of biodiversity conservation.[7] A 2017 study published in the journal *Nature Plants* estimates that botanic gardens today contain over 30% of all known plant species as *ex situ* conservation.[8] Many of these plants were collected in the context of the histories of colonialism that transverse a good number of botanic gardens around the world. Formerly a colonial botanical garden, Royal Botanic Gardens Kew, in England, UK, is now one of the greatest centers for plant biodiversity research, education, and conservation.[9] As host to a large number of tropical plants, some of which are extinct or near extinct in their ecosystems of origin, Kew is a prime example of the great responsibility that some botanic gardens carry in relation to *ex situ* conservation. Kew is also involved with *in situ* conservation, which it carries out in collaboration with botanic gardens around the world – many in former colonies of the British Empire.

A final example of the crucial roles that botanic gardens play in relation to plant biodiversity conservation is their involvement with Global and National Strategies for Plant Conservation. A select group of botanic gardens have participated very centrally in the development of these strategies. They seek to implement and materialize protective measures for plant and fungi life forms, thus bringing to fruition general conservation principles for plants stemming from the Convention on Biological Diversity. This achievement, the result of a monumental research and policy writing effort, has provided the world with clear guidelines,

1 Leaders in Conservation, "Leaders in Conservation: Botanic Gardens and Biodiversity in the 21st Century," https://leadersinconservation.wordpress.com.

2 Donald Rakow & Sharon Lee, "Western Botanical Gardens: History and Evolution," *Horticultural Reviews* 43 (2015): 269–310; Michael R. Dove, "Plants, Politics, and the Imagination Over the Past 500 Years in the Indo-Malay Region," *Current Anthropology* 60, Supplement 20 (2019): S309–S320.

3 Botanic Gardens Conservation International, https://www.bgci.org/.

4 Ibid.

5 For a review of the literature, see Katja Neves, "Tackling the invisibility of abeyant resistance to mainstream biodiversity conservation: Social movement theory and botanic garden agency," *Geoforum* 98 (2019): 254–63.

6 Botanic Gardens Conservation International, "Jardim Botânico do Faial," https://tools.bgci.org/garden.php?id=1465.

7 Katja Grötzner Neves, *Postnormal Conservation: Botanic Gardens and the Reordering of Biodiversity Governance* (State University New York Press, 2019).

8 Ross Mounce, Paul Smith & Samuel Brockington, "Ex situ conservation of plant diversity in the world's botanic gardens," *Nature Plants* 3 (2017): 795–802.

9 Botanic Gardens Conservation International, "Royal Botanic Gardens at Kew," https://tools.bgci.org/garden.php?id=766.

10 Neves, *Postnormal Conservation*.

11 BCGI "Growing the Social Role of Botanic Gardens," *BG Journal* 9, no. 1 (2012): 28–31; Katja Neves, "Lay Expertise and Botanical Science: A Case of Dynamic Interdependencies in Biodiversity Conservation," in E. Turnhout, W. Tuinstra & H. Willem (eds), *Environmental Expertise: Connecting Science, Policy, and Society* (Cambridge University Press, 2019), 200–209.

12 For a BGCI-produced animation that illustrates this shift, see www.youtube.com/watch?v=c_592TG6eQI.

directives, and measures for conservation worldwide. Many botanic gardens globally rely on these foundations to guide their missions and practices. Although much remains to be done, over the years the cumulative success of these efforts has been a game changer in plant biodiversity conservation.[10]

New Social Roles for Botanic Gardens

Over the past four decades a growing number of botanic gardens have enhanced their public education and public outreach programs. They have also been engaged with expanding their social roles in relation to the communities within which they are embedded and beyond.[11] In the Global North this partly stemmed from the recognition that the content and engagement strategies of "old" European and Northern American botanic gardens were designed mostly in alignment with the sensitivities and backgrounds of white, middle-to-upper-class visitors. As a result of this realization, many botanic gardens began working to shift their approaches towards greater inclusivity.[12] They have been striving to expand their content so as to reach much wider levels of diversity relating to class, ethnic-cultural affiliations, age, neurological diversity, and bodily ability. Some have even embraced content co-creation strategies that further dismantle divides between scientific and non-scientific forms of expertise.[13]

A large number of botanic gardens in the Global South boast rich histories of community engagement and play important social roles in the local, regional, and national settings of which they are part. Some are repurposed former colonial botanical gardens as is the case, for example, of the *Jardim Botânico do Rio de Janeiro* in Brazil. Many others were created in the context of the decolonization movements of the mid- to late-20th century and were developed specifically to serve socio-environmental goals. This includes, for example, ethnobotanic gardens designed to record, curate, and preserve local/traditional expertise on plant knowledge, such as the *Kawasan Wisata Pendidikan Lingkungan Hidup* (KWPLH) in East Kalimantan, Indonesia and/or national botanic gardens like the *Jardín Etnobotánico de Oaxaca* in Mexico whose mission it is to secure the protection of local endemic native plants.[14]

The Global South also hosts botanic gardens that were instituted within the parameters of European colonialism and have since been reinventing themselves as postcolonial biodiversity conservation organizations. These too fulfill important social roles, including the curation of local heritage, stewardship, and conservation of some of the planet's most biologically diverse ecosystems, as well as critically important leadership in bringing together diversity of knowledge practices from science to Indigenous knowledge. The Kirstenbosch National Botanical Garden of South Africa is an example of a botanic garden in the Global South that was specifically developed to preserve indigenous plants.[15] Located in lands once owned by Cecil Rhodes, this garden embraced a conservation directive in 1913, which precedes by decades the shift toward a conservation ethos noted in former colonial botanic gardens in the Global North, as well as the foundation of Botanic Gardens Conservation International.

The Kirstenbosch National Botanical Garden of South Africa also exemplifies the complex potential that botanic gardens have, despite their colonial legacies, to decolonize dominant Western epistemologies and associated plant production practices. In this context, Melanie Boehi's work reveals former colonial sites can be used to disrupt Western knowledge practice matrices so as to unveil hidden Indigenous narratives and understandings.[16] Doing so allows for the emergence of alternative conceptualizations of human-plant relations that are not imbued with Western fallacies of control. This can facilitate the emergence of more resilient multi-species arrangements within and beyond the botanical garden institution, with the related effect that the botanical garden institution engages with the governance of biodiversity conservation.[17]

Such a shift is critically important, especially considering the growing body of scholarship that has amply demonstrated that the effective mitigation of the world's most serious environmental problems, namely biodiversity loss and climate change, will require moving past the reductionism of Western ways of knowing and acting.[18] This, in turn, entails not only dismantling conceptual hierarchies between Western scientific knowledge practices and lay/Indigenous forms of expertise but also the associated structures and practices of coloniality that continue to erase and silence lay and Indigenous expertise.[19] Indeed, a large pool of scholarly studies have shown that lay and Indigenous expertise are particularly apt at recognizing and working sustainably with the holistic complexity of human/plant/animal embroilment.[20]

A final example of the important socio-environmental missions that botanic gardens have taken upon themselves in recent years is the matter of food security. With the price of food skyrocketing as a result of COVID-19 related supply chain disruptions, the invasion of Ukraine by Russia, and climate change induced agricultural losses, access to food has become even more difficult for populations who already faced food scarcity before these developments. For many in countries in North America and in Europe, food scarcity has reached scales not seen since the aftermath of WWII. This is especially concerning in regard to access to quality plant-based nutrition.

For some botanic gardens, engagement with matters of food security comes in the form of research within wider contexts of governmental strategies to adapt to climate change. DNA research is currently being carried out at botanic gardens to identify plant genes that are particularly well suited for specific climate conditions. The hope is that one day this research will facilitate the development of agricultural crops that are better adapted to a changed climate. However, not all solutions put forth by botanic gardens in relation to food security are based on techno-scientific solutions such as these. In some cases, botanic gardens can offer expertise with plant acclimatization. In fact, during the colonial period of botanical garden history one of the key functions of botanic gardens was precisely to make possible the adaptation of plant species within the domain of imperial nation-states. Moreover, in recent years some botanic gardens have followed decolonial approaches to this practice, working alongside Indigenous communities who, over the course of centuries, developed highly effective practices to mitigate the effects of climate oscillations.

In parallel to these developments, a group of botanic gardens within the BGCI network developed and implemented a series of initiatives that address issues of food security in urban centers. Called *Communities in Nature*, these projects sought to achieve social and environmental goals through gardening activities.[21] As a collective spread throughout a network of botanic gardens around the world, staff working within the context of botanical garden public outreach operate in ways that resemble those characteristic of social movements dedicated to the attainment of socio-environmental goals.[22] This philosophy has been steadily growing in recognition and is increasingly present in botanical garden design. The recently created Portland Botanic Gardens in Oregon, USA, illustrates this trend, which has been solidifying in many areas of the globe in recent decades.[23]

Conclusion

A common public perception of botanic gardens is that they are quaint institutions for general audiences to enjoy as green spaces of leisure, and for scientists to pursue botanical research. While botanic gardens are indeed green havens in urban spaces that can serve important therapeutic purposes, and remain crucial sites for the study of plants and their applications, the current relevance of the botanic garden institution far surpasses the parameters of these roles. Botanic gardens around the world are productively engaged with tackling biodiversity on a global scale, increasing food security for the world's current and future populations, and decolonizing the very kinds of knowledge and practices that contributed to our current global woes. Far from irrelevant, and thanks to their great institutional resiliency and capacity to reinvent themselves at different historical junctures, botanic gardens are important agents in the pursuit of socio-environmentally sustainable futures for all.

13 Katja Neves, "Lay Expertise, Botanical Science, and Botanic Gardens as 'Contact Zones'," *Oxford Research Encyclopedia of Environmental Science* (November 29, 2021). Also see Neves, "Lay Expertise and Botanical Science" and "Tackling the invisibility of abeyant resistance to mainstream biodiversity conservation."

14 KWPLH, "About Us: A Greener Future Through Education," https://www.beruangmadu.org/kwplh/about-us/; Jardín Etnobotánico de Oaxaca, https://jardinoaxaca.mx.

15 Melanie Boehi, "Radical Stories in the Kirstenbosch National Botanical Garden: Emergent Ecologies' Challenges to Colonial Narratives and Western Epistemologies," *Environmental Humanities* 13, no. 1 (2021): 66–92.

16 Ibid.

17 Katja Neves, "Urban Botanical Gardens and the Aesthetics of Ecological Learning: A Theoretical Discussion and Preliminary Insights from Montreal's Botanical Garden," *Anthropologica* 51, no. 1 (2009): 145–57; Neves, *Postnormal Conservation*.

18 See, e.g., Ilisapeci Lyons, et al. "Protecting what is left after colonisation: embedding climate adaptation planning in traditional owner narratives," *Geographical Research* 58, no. 1 (2020): 34–48; Danielle Emma Johnson, Meg Parsons & Karen T. Fisher, "Indigenous climate change adaptation: New directions for emerging scholarship," *Environment and Planning E: Nature and Space* 5 (2021): 1541–78; Martin Mahony & Mike Hume, "Epistemic Geographies of Climate Change: Science, space and politics," *Progress in Human Geography* 42, no. 3 (2018): 395–424; Allan Rarai, et. al., "Situating Climate Change Adaptation Within Plural Worlds: The Role of Indigenous and Local Knowledge in Pentecost Island, Vanuatu," *Environment and Planning E: Nature and Space* 5, no. 4 (2022): 1–43.

19 W.D. Mignolo, "Epistemic Disobedience, Independent Thought and Decolonial Freedom," *Theory, Culture & Society* 26, no. 7-8 (2009): 1–23.

20 See note 13.

21 BCGI, "Communities in Nature," https://www.bgci.org/our-work/projects-and-case-studies/communities-in-nature-growing-the-social-role-of-botanic-gardens/.

22 Neves, "Lay Expertise and Botanical Science."

23 Portland Botanical Gardens, "First Nations Gardens," https://portlandbg.org/visit/grounds/demonstration-gardens/first-nations-gardens/.

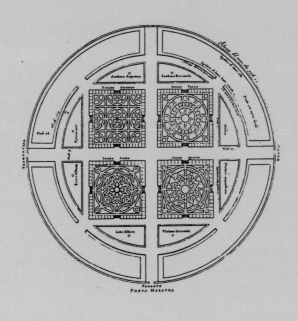

SONJA DÜMPELMANN
SPIRALING DIVERSITY AND BLANK SPOTS
IN A 19TH-CENTURY UTOPIAN BOTANIC GARDEN

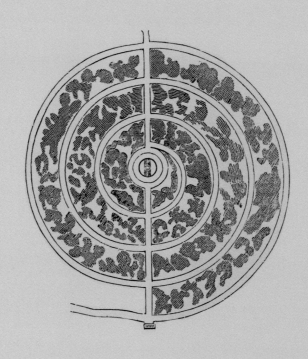

Sonja Dümpelmann is a landscape historian and professor at Ludwig-Maximilians-Universität in Munich, where she holds the Chair of Environmental Humanities and codirects the Rachel Carson Center. She is the author and editor of several books, most recently of *Landscapes for Sport* (Dumbarton Oaks, 2022), and of the award-winning *Seeing Trees: A History of Street Trees in New York City and Berlin* (Yale University Press, 2019). She lectures internationally and has served as Senior Fellow in Garden and Landscape Studies at the Dumbarton Oaks Research Library and Collection, Washington, DC.

+ HISTORY, DESIGN

On December 17, 1811, 28-year-old John Claudius Loudon addressed the Linnean Society in London with an idea and a plea. Already a recognized landscape gardener inducted into this group of amateur naturalists five years earlier, Loudon proposed to establish a National Garden. By this he meant a new type of botanic garden in the metropolis. Despite the well-established botanic gardens in Kew, Cambridge, and Oxford, and the more recent one in Liverpool, Loudon considered his country to be lagging in this area. He was especially disappointed with the botanic gardens' aesthetic and limited use. The National Garden therefore was to be not only a scientific institution serving botanists but a "living museum" and educational venue serving the public at large for whom it would also function as a park.[1] Furthermore, it was to become "the first school for practical gardeners in the world."[2] Besides a collection of individual plant species and cultivars, the garden would show different garden styles from various parts of the world as well as agricultural and horticultural practices for the cultivation of food plants and flowers. Although Loudon admitted on later occasions that science could be allowed to trump art in botanic gardens, this was not his ideal.[3] The ideal was the combination of both.[4] Loudon's appeal to the Society therefore included a description of his design vision for the garden. Various plant arrangements would be positioned along an Archimedean spiral that centered on a large circular glasshouse accommodating "greenhouse plants" and "stove exotics" from warmer climates.[5] Plants were to be arranged using a variety of ordering systems, including the alphabet, the Linnaean taxonomy, and the "natural method" developed by Bernard and Antoine Laurent de Jussieu.[6] In addition, Loudon envisioned the National Garden to provide examples of picturesque arrangements à la Uvedale Price and Richard Payne Knight, and of seasonal plantings offering flowering scenes for each month of the year.[7]

Spiraling into the Future

Loudon's 1811 vision for a National Garden was an elaborate, informed mixture of old and new botanic classifications and landscape aesthetics. The design illustrating this vision one year later in his *Hints on the Formation of Gardens and Pleasure Grounds* was aimed at a wider audience and therefore stripped of many smaller details and ideas. However, Loudon retained the general concept and form based upon the Archimedean

SPIRALING DIVERSITY AND BLANK SPOTS

Above: Loudon's utopian botanic garden. John Claudius Loudon, *Hints on the Formation of Gardens and Pleasure Grounds* (Gale, Curtis & Fenner, 1813) plate 17.

Next: John Claudius Loudon, *An Encyclopedia of Plants* (Longman, Rees, Orme, Brown & Green, 1829), 861, 868.

spiral. In what was now entitled a "Design for a Botanic Garden arranged so as to combine elegance & picturesque effect with botanical order & accuracy," the clockwise coil again centered on a large glasshouse for stove and greenhouse plants. It was enclosed by a sheltering belt of mostly evergreen shrubs intermixed with fruit trees that would provide colorful blossoms in the spring.[8] The spiral pathway gave expression to a confident, deep-seated belief in human progress. It could be extended endlessly, marking time into the future.

In various cultures around the world since paleolithic times, the spiral has been symbolic of life and growth, creation and becoming, evolution and development, as well as knowledge, perseverance, and perpetuity. While ubiquitous in nature, it is also one of the oldest abstract geometric motifs used in art and architecture, appearing in stone engravings, temples, weaponry, equestrian items, household artifacts, as well as in ceremonial and personal objects such as jewelry.[9] Since neolithic times spirals have been an organizing principle in placemaking and architectural organization, for example in mythic and real labyrinths, including the corridors leading to Egyptian burial chambers created during the New Kingdom.[10] In Egyptian decorative art, the spiral also appeared as an abstract vegetal ornament. As art historian Alois Riegl sought to show at the height of the arts-and-crafts movement in the late 19th century, over time the spiral evolved from geometric pattern to more explicit ornamental representations of plant tendrils, a development that he observed on the basis of Egyptian friezes, Melian, Rhodian, and Chalcidian vases as well as subsequent Roman wall frescos.[11] Thus, in human culture, the spiral has for a long time also been directly associated with plant life. Even if plants' spiral growth had been observed by both artists and naturalists long before, in the 1820s and 1830s during Loudon's lifetime, the topic attracted the renewed attention of several botanists and naturalists including the German romantic writer Johann Wolfgang von Goethe.[12] He assumed "in vegetation a general spiral tendency," which, together with "a vertical force" was responsible for plants' metamorphosis.[13] Whereas the vertical system was the durable, solidifying, more permanent plant fibers, Goethe saw the spiral system as "the developmental, reproductive, and nourishing element."[14] Together both systems would produce the most perfect vegetation if they were in complete equilibrium.[15] If Goethe described plants' life force as a dialectic of spiral and vertical tendencies, more recently in 1982, structuralist semiotician Roland Barthes saw the spiral form itself as "dialectical;" as "governing the dialectic of the old and new," and as "a circle distended into infinity, where things recur but at another level… in difference." On a spiral, he observed, "nothing is first yet everything is new."[16]

These symbolic meanings can also be attributed to Loudon's use of the spiral. In his utopian botanic garden design, the coil provided a stable framework within which transformation could occur. In other words, the design built upon tensions

1 John Claudius Loudon, "Hints for a National Garden," 3, John Claudius Loudon papers, SP 716, Linnean Society Archives, London.

2 Ibid., 15.

3 John Claudius Loudon, *Encyclopaedia of Gardening* (Longman, Hurst, Rees, Orme & Brown, 1822), 1,189–90.

4 Loudon, "Hints for a National Garden," 20.

5 Ibid., footnote on separate sheet between pages 9 and 10, 20.

6 Loudon suggested the "natural method" especially for ordering plants from warmer climates in the glasshouses. This choice corresponded to the findings of botanist Michel Adanson who, while in Senegal in the mid-18th century, questioned the applicability of Linnaeus's system to the plants and animals in this part of the world. Based upon earlier critiques voiced by Buffon, he promoted a "natural method" built upon all plant characteristics rather than only those of sexual reproduction. See Richard Drayton, *Nature's Government: Science, Imperial Britain, and the "Improvement" of the World* (Yale University Press, 2000), 19.

7 Loudon, "Hints for a National Garden," 11.

8 John Claudius Loudon, *Hints on the Formation of Gardens and Pleasure Grounds* (Gale, Curtis & Fenner, 1813), 30, plate XVII.

9 For example, spiral patterns were carved into the stone covering a neolithic burial chamber at Newgrange, Ireland and the stone walls of the neolithic Tarxian temples in Malta; they were used as ornaments on Japanese clay figures and pottery of the Jomon period (14,000–400 BCE); on vessels, chairs, columns, as well as walls and ceilings of Egyptian burial chambers of the Middle and New Kingdoms; on Scythian golden neckpieces (900–100 BCE), and on Maori wood and stone carvings. Gertrud Thausing, "Das Symbol der Spirale im alten Ägypten," *Wiener Zeitschrift fur die Kunde des Morgenlandes* 56 (1960): 241–49; Kinko Tsuji & Stefan C. Müller, eds., *Spirals and Vortices In Culture, Nature, and Science* (Springer, 2019).

10 Thausing, "Das Symbol der Spirale im alten Ägypten."

11 Alois Riegl, "Chapter 3. The Introduction of Vegetal Ornament and the Development of the Ornamental Plant Tendril," *Problems of Style: Foundations for a History of Ornament*, transl. by Evelyn Kain, ed. by David Castriota (Princeton University Press, 2018 [1992]), 48-228. Also see David Castriota, "Annotator's Introduction and Acknowledgements," in David Castriota (ed.), *Problems of Style: Foundations for a History of Ornament*, transl. by Evelyn Kain (Princeton University Press, 1992), xxv–xxxiii.

12 William M. Montgomery, "The Origins of the Spiral Theory of Phyllotaxis," *Journal of the History of Biology* 3, no. 2 (1970): 299–323.

13 Johann Wolfgang von Goethe, "The Spiral Tendency," in *Goethe's Botanical Writings*, transl. by Bertha Mueller (University of Hawaii Press, 1952), 129.

14 Ibid., 132.

15 Ibid., 131–32.

between the relative fixity of the regular spiral plan, and the malleable irregular outlines of the individual planting beds encompassed within it; between knowledge and certainty on the one hand, and ignorance and uncertainty on the other; between scientific classification system as well as cosmic form and spiritual meaning. The spiral connected the plant and human scales with the nonhuman scales of deep time. The garden's overall organization could in more general terms also be compared to that of the earth's solar system. The glasshouse with its stove emanating warmth and accommodating plants from the warmest and sunniest climates was positioned in the middle of plant collections that orbited around the center along the spiraling path. A straight cross-sectional pathway connected the two entrances lying on opposite sides of the spiral with the central glasshouse, thus providing fast access to all parts of the garden. The design therefore allowed for leisurely wandering along the spiral, and for deliberate walking along the straight route crossing the spiral's turns and leading to the center. The garden accommodated movement combining or alternating between pleasure and purpose.

Long before Loudon's design, botanic gardens had been based upon a circular if not spiraling organization and layout. In one of the earliest, the famous garden in Padova founded in 1545, a circle of planting beds enclosed a square divided into quadrants symbolizing the four corners of the world known at the time: Europe, Asia, Africa, and America. Mirroring the aesthetics of pleasure gardens at the time, the quadrants were laid out as parterres consisting of intricate geometric patterns. They were formed by planting beds for individual plant families or species. The planters' conceit was that the botanic collection established for medical and plant studies would contain the world in a garden, an idea that for many also had a spiritual meaning. As a miniature world of creation, early botanic gardens were also thought to be a means to understand God.[17] Given the acclaim of Padova's garden, most botanic gardens established in the following centuries were divided into quadrants. Some designers also resorted

16 Roland Barthes, *The Responsibility of Forms: Critical Essays on Music, Art, and Representation* (Hill & Wang, 1985), 218-19.

17 See Drayton, *Nature's Government*, 9-12.

18 On Petersen, see Denis Diagre-Vanderpelen, *The Botanic Garden of Brussels (1826-1912): Reflection of a Changing Nation* (National Botanic Garden of Belgium, 2011), 26; Xavier Duquenne, "Drie Duitse tuinarchitecten (Charles-Henri Petersen 1792-1859, Louis Fuchs 1818-1904, Édouard Keilig 1827-1895)," *Historische Woonsteden & Tuinen* 157, no. 1 (2008): 19-20.

19 Drayton, *Nature's Government*, 20.

20 See for the sprawl at Leyden botanic garden, Drayton, *Nature's Government*, 24.

21 Loudon, *Hints on the Formation of Gardens and Pleasure Grounds*, 30, plate XVII. Loudon re-published this design some years later in the first edition of his *Encyclopaedia of Gardening* (1822) to illustrate a "private botanic garden," (p. 908), offering further descriptive comments.

22 Alfred W. Crosby, *The Columbian Exchange: Biological and Cultural Consequences of 1492* (Greenwood Press, 1972); Drayton, *Nature's Government*, 271.

23 See John Claudius Loudon, *Encyclopaedia of Gardening* (London: Longman, Rees, Orme, Brown & Green, 1825), 107.

24 Ibid.

25 Ibid., 107-8.

to a circular form, for example German garden architect Carl Heinrich (Charles-Henri) Petersen at the Brussels botanic garden that opened in 1826.[18] However, already beginning around 1650, not least due to colonial expansion, botanists realized that their hope of collecting all plants and medicinal knowledge in one garden—no matter whether round or square—was in vain.[19] Gardeners took pains to accommodate the newly introduced plants but could seldom prevent their unorganized sprawl beyond the initial geometric layouts of squares and circles.[20]

Loudon's design addressed this challenge. In contrast to the more static form of the whole, closed circle in Padova and Brussels, and to the confining quadrangular layouts of many other gardens, Loudon's spiral was dynamic and open. It could be extended incrementally and indefinitely, accommodating transformation and growth. Similarly, the planting beds along Loudon's spiral were "undefined." They had irregular, amorphous, and differently sized shapes. In these ways, Loudon explicitly accommodated the idea of uncertainty within a design that otherwise gave expression to an unabashed belief in future progress.

Designing Botanic Blank Spots

The vagaries and uncertainties of future developments in botanical science led Loudon not only to draw amorphous planting bed outlines, but to redirect attention to the shapeless blank spots, or "white" spaces in-between. These were the areas meant to accommodate future plant discoveries and developments in botanical science. Intending to "combine elegance & picturesque effect with botanical order & accuracy," the botanic garden was to comprise a collection of plants growing in Britain and arranged according to the Linnaean system – that is, the visual recognition of plants' reproductive parts. Each of the 24 classes and their orders was allotted an irregularly shaped planting bed along the spiraling pathway. The spaces in between the planting beds were laid out with smooth turf so that the beds could easily be

26 John Claudius Loudon, *Arboretum et Fruticetum Britannicum; or, The Trees and Shrubs of Britain*, vol. I (Longman, Orme, Brown, Green, and Longmans, 1838), 191.

27 Ibid., 3–4, 191.

28 For this shift that occurred also more generally in botanic science as well as the relationship to aesthetics in landscape gardening, see Therese O'Malley, "Art and Science in the Design of Botanic Gardens, 1730–1830," in *Garden History: Issues, Approaches, Methods*, edited by John Dixon Hunt (Dumbarton Oaks Research Library and Collection, 1992), 279–302.

29 See Ibid., 288.

30 Loudon, *Encyclopaedia of Gardening* (1822), 908–9.

31 See Michel Foucault, *The Order of Things* (New York: Vintage Books, 1994 [1970]),126–27.

32 Loudon, *Encyclopaedia of Gardening* (1822), 1225; Melanie Simo, *Loudon and the Landscape: From Country Seat to Metropolis* (Yale University Press, 1988), 114, 264.

33 Loudon, *Arboretum et Fruticetum Britannicum*, vii, 8, 84, 118–19.

34 John Claudius Loudon, "Catalogue of Works on Gardening, Agriculture, Botany...," *The Gardener's Magazine* 1 (1826): 318; John Claudius Loudon, "Garden Botany," *The Gardener's Magazine* 1 (1826): 55–56.

35 Loudon, "Catalogue of Works on Gardening, Agriculture, Botany...," 318.

enlarged if new species were introduced and needed to be added.[21] These blank, or rather, "green" spots therefore literally gave space to a pending state of ignorance in botanical science and to its unknown future. Although the planting beds could be both expanded and contracted, at a time preceding today's overwhelming experience of biodiversity loss, Loudon only anticipated the discovery of new species, which implied an increase of plants that needed to be accommodated in botanic gardens. He was oblivious to the extinction of species that was already occurring, paradoxically, due to the same colonial exploitation that introduced new plants in the mother country in the first place.

In the early 19th century, an unprecedented belief in technological and scientific progress as well as territorial expansion reigned, fueled by widening imperial interests. Britain, besides Portugal, Spain, the Netherlands, and France, had already begun to establish colonial botanic gardens in the 17th century, to facilitate what historians Alfred Crosby and Richard Drayton have respectively called the Columbian and The Royal Exchange.[22] As many of his contemporaries, Loudon did not question imperialism, its social and political order, and saw in horticulture and gardens civilizing agents used by colonizers to quite literally acculturate "barbarous countries" and their "rude society."[23] He posited that "The influence of gardening comforts, together with instruction, on uncivilised countries...cannot be foreseen."[24] African deserts might in the future "be watered and cultivated...effect[ing] a material change in the climate," leading "millions of human beings [to] live and exert their energies where civilised man at present scarcely dares to tread."[25] Like his fellow botanists and contemporary Englishmen more generally, Loudon saw no conflict in using scientific knowledge to enrich the Empire and in promoting a cosmopolitan botanic utopia based upon "free and universal exchange."[26] It did not appear to him that his ideal of a "comparatively equalised" civilization inhabiting a globe with plants distributed throughout all regions where the climate allowed them to thrive was not built on mutual benefit.[27] In fact, the countries where many of the plants originated that would be populating the blank spots along the spiral in Loudon's prototypical botanic garden were themselves treated as blank spots on the map of an increasingly global world. In the eyes of many colonizers these blank spots were territories to be exploited as well as "civilized" and "improved." Assembling and cultivating indigenous plants in the colonial botanic gardens, and introducing them in the mother countries was one important step in this despotic endeavor; their naming and ordering according to European classification systems and ideas of beauty another.

In his initial notes for the National Garden Loudon proposed the use of both Linnaeus's and Jussieu's systems, depending on the type of planting and its intended use. His 1812 design suggested only the Linnaean order. A few years later, however, Loudon had moved away from the Linnaean taxonomy although it served students well.[28] Instead, he favored Jussieu's *méthode naturelle*. Bernard de Jussieu and his nephew Antoine Laurent had developed this classification beginning in the 1750s by grouping plants according to the number of their formal similarities. Given that order here was considered to emerge out of nature itself, the system would, Loudon thought, also create more harmonious and "natural" plant arrangements, and therefore appeal more easily to a broader public. This had been shown in the Parisian botanic garden where Antoine de Jussieu had ordered plants according to his natural method and habitat types.[29] When Loudon republished his utopian botanic garden design in the first edition of the *Encyclopaedia of Gardening* in 1822, he therefore espoused Jussieu's system proposing that the planting beds could also include artificial bogs, ponds, and springs to reflect different habitat types.[30] Jussieu's "natural method" indeed facilitated an aesthetic similar to the picturesque ideal of the time. The classification system therefore offered itself to infuse scientific spaces with contemporary landscape art and vice versa, to bring science into the landscape garden. It could support Loudon's aspiration of a scientific reform of landscape aesthetics through diversifying the species palette used in the creation of picturesque scenes. By doing so, it also gave expression to the changing perception of humans' relationship to nonhuman nature. Once considered static and unchangeable, natural phenomena now appeared more mutable.[31]

Growing the Spiral

Loudon's National Garden and utopian design were never realized. But some of the ideas he developed, voiced, and illustrated in this regard informed the Hackney Botanic Nursery of C. Loddiges & Sons, and the Derby Arboretum, designed by Loudon and opened to the public in 1840. The enthusiasm with which Loudon reported in the 1820s on the work of George and William Loddiges at their Hackney Botanic Nursery was perhaps partly self-congratulatory. Despite a lack of direct evidence there is reason to believe Loudon's botanic garden design inspired the brothers to lay out their arboretum in 1818 along a spiraling pathway. At the time, the nursery was also erecting its camellia house–part of an early world-renowned glasshouse ensemble–following Loudon's innovative design with iron-framed curvilinear glazing.[32] Loudon in turn used the Loddiges' tree collection as the basis for his work on *The Trees and Shrubs of Britain* published in 1838.[33]

Encompassing 2,664 hardy tree and shrub species, many introduced from America, the Loddiges' arboretum was in Loudon's estimation beyond worldwide comparison and its "value to the country...incalculable."[34] The brothers had "done more than all the royal and botanic gardens put together."[35] Not less remarkable than the number of species was their arrangement and the arboretum's design. Loudon published a plan in the second and all following editions of his *Encyclopaedia of Gardening*.[36] He described the design's defining spiraling path as "forming a scroll like the Ionic volute, extending over a space of upwards of seven acres."[37] Planted along the right side of the revolving gravel walk, tree and shrub species and their varieties were ordered alphabetically so they could be "examined with ease, and compared with [their] congeners at any time of the year."[38] Along the left side, the nursery's rose collection was planted followed by herbaceous plants.[39] In the arboretum's center 10 concentric zones were dedicated to "peat earth plants," again ordered following the alphabetical order of their names.[40]

In short, to Loudon, the Loddiges' arboretum appeared exemplary in many ways, and he encouraged gardeners as well as ladies of the landed gentry, especially mothers, to visit it. All good education, Loudon believed, began with mothers teaching their children. "What a paradise this island will become," he suggested, once education in arboriculture, horticulture, and landscape gardening established the groundwork for the use of "all the trees and shrubs in the world which will grow in temperate climates."[41] The point was, he surmised, to create grandiose landscape scenes such as the ones seen in the Niagara panorama on show in London at the time of his writing.[42] Believing that high and equal education was the birthright of every individual living in a civilized state, no matter their class, Loudon saw in botanic gardens a potential means of public education.[43] After most of his early texts had been produced lavishly and had been dedicated to the royalty, nobility, and gentry, beginning in the 1820s, his works—often dealing with horticultural labor and practices—shifted to addressing self-educated young gardeners, craftsmen, and middle-class householders.[44] Over the course of his career Loudon also exhibited an increasing awareness of the working conditions of laborers and their individual fates, but although he held some radical beliefs he remained caught in his own middle-class perspective and never publicly supported political movements such as Radicalism or Chartism.[45]

Labeling Plants

Loudon's project of turning botanic gardens into public-facing institutions that were at once living museums and parks required that plants be given the space and time to grow into their characteristic shape. They also needed to be labeled conspicuously yet unobtrusively. Reading plant names was crucial "for ladies to acquire a practical knowledge of botany," Loudon argued.[46] Strong cast-iron tallies with English plant names complemented with "Linnaean and Jussieuean class, native country, and time of flowering" would facilitate visitors' "peripatetic study of plants" as long as letters were large enough to be legible without stooping.[47] Nevertheless, at the time labels were lacking in Kew's arboretum and at the London Horticultural Society's garden in Chiswick. In contrast, Loudon attested, that good examples of labeling existed in some private gardens, for example in Mr. Boursault's garden in Paris, located on rue Blanche, and in Diana Beaumont's conservatory at Bretton Hall where glazed china tickets were used.[48] As new materials became available in the 19th century, number and name labels became an important discussion topic in horticultural circles. In 1822, Loudon listed 10 basic types of plant labels in his *Encyclopaedia*, and materials and labeling conventions were changing fast.[49] One of the oldest

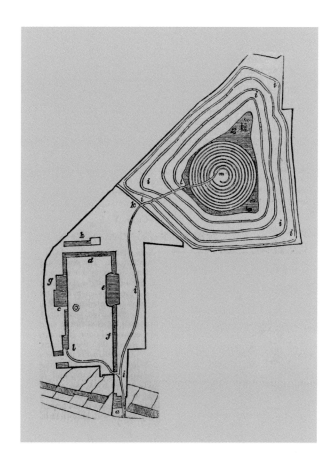

Above: Hackney Botanic Nursery of C. Loddiges & Sons with its arboretum organized along a spiraling pathway. John Claudius Loudon, *Encyclopaedia of Gardening* (Longman, Rees, Orme, Brown & Green, 1824), 1035.

numbering systems used by botanists, but that was less practical for public display due to its coded notched marks, was the common tally, or number-stick.[50] At the Loddiges' nursery the direct designation of plant names appeared more practical and customer-friendly. However, like the plants themselves, any type of label had to withstand the vagaries of weather and climate. In the 1830s the nursery had to repaint many plant names on porcelain tallies with thick black oil paint after the previously used thin paint had dissolved. The latter had replaced black-lead pencil, which was even less clear and durable unless varnished.[51] The nursery also used "number-bricks" in its more permanent tree and herbaceous arrangements. These bricks were set endways and obliquely into the soil and the number designating the respective species was painted on a black or white ground.[52]

Loudon chose this labeling method for the tree and shrub species exhibited at the Derby arboretum. Sponsored by industrialist Joseph Strutt, Loudon designed the arboretum grounds in 1839. Finally, here was an opportunity for him to realize some of his design ideas, pairing education with recreation. Limited funds prevented the implementation of a complete botanic garden and because the unseemly site did not offer itself to the display of herbaceous plants, Loudon's penchant for trees and shrubs could come into its own unhindered. After all, as he pointed out, woody plants also attracted songbirds, and trees offered themselves for educational intent due to their size and characteristics recognizable by perambulating visitors in motion.[53] Every woody plant in the grounds belonged to a different species, and they were arranged following the natural method. To ensure that visitors of the new pleasure ground could learn about each species' name, history, uses, folklore, and seasonal characteristics, plants were labeled and referred to in a printed catalog. In addition, the abridged version of Loudon's comprehensive work on *The Trees and Shrubs of Britain* was available for consultation in the arboretum's entrance lodges.[54] To facilitate an educational and pleasurable experience Loudon initiated ground modulation and required that all trees and shrubs be planted on small "hills." In this way Loudon not only created variable spaces, but elevated trees and shrubs to exhibition objects, furthering plant and arboricultural knowledge. To illustrate their natural habit, trees were to grow without pruning. To demonstrate correct planting practice Loudon made sure that trees' root flares remained exposed; and to model mulching practice and preserve soil moisture, he advised that fallen leaves be used to cover the areas surrounding tree trunks.[55]

The Derby arboretum was Loudon's testing ground for diversifying British designers' and gardeners' plant palette and for producing the gardenesque, an aesthetic he had developed beginning in the early 19th century and first presented in writing in 1832.[56] When seen from a distance, trees should appear as if planted in connected groups, not dissimilar to the picturesque principle. Up close, however, they were to take shape as individual specimens so their unique form could be perceived and studied. At the arboretum this meant that if branches from different species became entangled and if trees grew taller than 40 or 50 feet these individuals were to be removed. With this drastic measure Loudon also aspired to maintain and enhance species diversity, making sure that some trees would not crowd out others, and that newly discovered and introduced species could be exhibited on what was limited ground.[57] The site was too confined to allow for the blank spots of his botanic utopia, after all. Loudon's gardenesque aesthetic found distribution in handbooks but only limited emulation in design practice. However, his argument that botanic gardens were a matter of both art and science was repeated by future landscape gardeners and architects who took botanic gardens into their portfolio.

Today, the conservation of endangered species has become an additional mission of many botanic gardens besides research in the plant sciences, botanic education, and the provision of park space. New botanic gardens build upon the realization that no garden can accommodate samples of all living plant species and that the number of known existing species is changeable, especially in the current time of accelerated species extinction. As at the Jardí Botànic in Barcelona, Spain (1999) and the Jardin Botanique de la Bastide in Bordeaux, France (2003), new gardens' plant exhibits and designs are often attuned to the respective climate zones and regions they are situated within, building upon the advances of plant ecology (and taxonomy). However, for all scientific accomplishments, where do we stand on Loudon's spiral? Conservation botanists observing plant extinction tell us that while current elimination rates are unknown, "we are already in the bottleneck of this extinction."[58] As the consequences of the 19th-century idea of progress and botany's entanglements in empire-building are increasingly being exposed, Loudon's spiral and "blank spots" have remained surprisingly current. Although the amorphous planting beds would today be contracting rather than expanding despite new plant discoveries, only more and better knowledge inspiring action can help get us through the bottleneck.

36 See John Claudius Loudon, *Encyclopaedia of Gardening* (Longman, Rees, Orme, Brown & Green, 1824), 1035, and all following editions.

37 Loudon, *Arboretum et Fruticetum Britannicum*, 130.

38 Ibid.; Loudon, "Catalogue of Works on Gardening, Agriculture, Botany....," 318.

39 Loudon, *Encyclopaedia of Gardening* (1825), 1035.

40 Loudon, *Arboretum et Fruticetum Britannicum*, 130.

41 John Claudius Loudon, "Calls at the London Nurseries, and other Suburban Gardens," *The Gardener's Magazine* 9 (1833): 468.

42 Loudon most probably referred to the landscape of Niagara painted by Robert Burford, on show at the panorama at Leicester Square in London in 1833. See Robert Burford, *Description of a View of the Falls of Niagara* (Brettell, 1833).

43 Simo, *Loudon and the Landscape*, 11.

44 Ibid., 9.

45 Ibid., 11, 12.

46 Loudon, "Horticultural Society and Garden," *The Gardener's Magazine* 1 (1826): 344–45.

47 John Claudius Loudon, *Encyclopaedia of Gardening* (London: Longman, Hurst, Rees, Orme & Brown, 1822), 1,190.

48 John Claudius Loudon, "Horticultural Society and Garden."

49 Loudon, *Encyclopaedia of Gardening* (1822), 325–27.

50 Loudon, *Encyclopaedia of Gardening* (1825), 280–81.

51 John Claudius Loudon, "Calls at the London Nurseries, and other Suburban Gardens," *The Gardener's Magazine* 9 (1833): 468.

52 Loudon, *Encyclopaedia of Gardening* (1825), 282.

53 John Claudius Loudon, *The Derby Arboretum* (Longman, Orme, Brown, Green & Longman, 1840), 72.

54 Ibid., 81–82.

55 Ibid., 72, 73, 80.

56 J.C.L. "Preface," *The Gardener's Magazine* (Longman, Rees, Orme, Brown, Green & Longman, 1832): iii-iv (iv); John Claudius Loudon, "Reviews," *The Gardener's Magazine* (Longman, Rees, Orme, Brown, Green & Longman, 1832): 701; Simo, *Loudon and the Landscape*, 87, 169–71.

57 Loudon, *The Derby Arboretum*, 80–81.

58 W. John Kress & Gary A. Krupnick, "Lords of the Biosphere: Plant Winners and Losers in the Anthropocene," *Plants People Planet* 4 (2022): 350–66; Eric W. Sanderson, Joseph Walston & John G. Robinson, "From Bottleneck to Breakthrough: Urbanization and the Future of Biodiversity Conservation," *BioScience* 68, no. 6 (2018): 412–26.

BONNIE-KATE WALKER

GARDEN OF RELATION: DRAWING THE CLIMATIC INTELLIGENCE OF PLANTS

Bonnie-Kate Walker is a research associate at the Chair of Being Alive at ETH Zürich, where she contributes to teaching and research on regenerative practices and drawing languages for living systems. She is a co-founder of the landscape design collective Office of Living Things, whose work centers social justice, relationships, and long-term ways of thinking about land. She received her MLA from the University of Virginia in 2017 and has since practiced in New York City and Zürich.

✚ DESIGN, HORTICULTURE

Anafi is a small Cycladic island in the Aegean Sea, where there are almost no trees, and the landscape is characterized by its extreme climatic condition: intense long days of sun and heat in summer, powerful winds, poor soil, and erosion. In a lecture for gardeners in 2018, nursery grower and gardener Olivier Filippi describes his work in Mediterranean conditions like these as "gardening on the moon."[1] I went to Anafi one summer in early July. The plants all appeared to be of similar color and shape: variations of a gray and brown in huddled, round masses, distributed around the baked slopes of the island. But behind this superficial monotony are powerful strategies that enable the plant's survival and, ultimately, lend to these plants a place-based beauty – characteristics that emerge from extreme constraints. The ground tapestry that characterizes the Cycladic landscape is formed by the bulbous topography of cushion plant morphology, the woody and sinuous texture of the plant's perennial parts that allow it to resprout after fire or drought, and the tiny, hairy leaves that retain condensation to slow evapotranspiration. The methods by which a plant expresses its climatic origin can be strange and beautiful, lending to a fertile design territory while hinting at some helpful measures for a hotter, drier future.

Plants have a climatic intelligence written into their morphology, and even their genetics. Unable to move quickly, put on a jacket, or turn on an air conditioner, plants must endure gradual and abrupt changes in climate and weather, making them the most responsive beings on earth. As a landscape architect, I am interested in a close reading of the climatic intelligence of plants for two reasons: one, it is amazing; and two, because landscape architects play an obvious role in designing more complex,

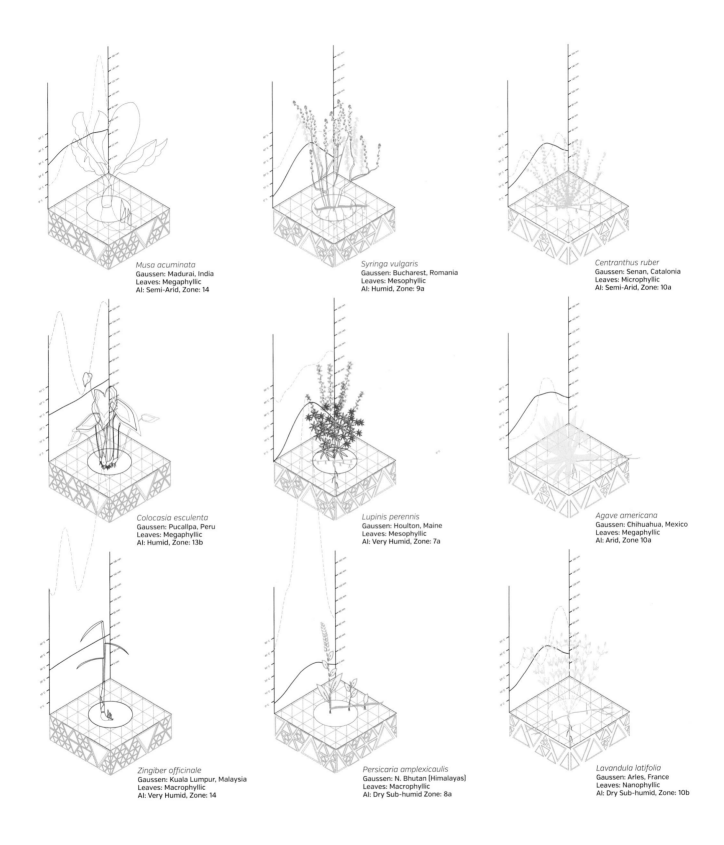

climate-responsive assemblages and in cultivating the demand for more specialized plants that fill those ecological niches. It is common to make choices about plants based on a variety of aesthetic parameters—color, texture, size, and bloom time—but more specificity and creativity with climatic criteria would yield better planting design. In the era of climate upheaval within which we are working, we should be inspired by plants that have learned to live in the broad range of climatic situations on the planet. However, there are some barriers to designing with this potential. One issue is the sourcing and availability of a diverse range of plants that could work in a specific climatic condition. But more importantly, landscape architects are not typically using adequate tools to approximate a plant's climatic suitability, particularly in terms of water. In this piece, I want first to explore the work of small nursery growers and researchers to better understand the climate-plant relationship and make that knowledge (and those plants) available to the public. Secondly, I want to share drawing tools that can help to integrate that thinking into the discipline of landscape architecture by adding the parameter of water to the USDA Plant Hardiness Zones classification. Inspired by the writings of thinkers like Gilles Clément and Édouard Glissant, it is my hope that this work ultimately contributes a more relational understanding of sites in a planetary context, and a more flexible yet precise approach to planting for future climates.

Plants that thrive in arid climates hold some keys to maintaining biodiversity in the future scenarios in store for much of the planet. Many of them tolerate drought by investing in their root system. Several species of the genus *Cistus*, for example, have a double root system that allows the plant in the early stages of its life to develop an extremely deep tap root at the expense of investing in the growth of its aboveground parts. The *Cistus* then grows an extensive superficial network of fibrous roots that soak up even small quantities of rainwater from the surface when it arrives, while using the taproot to access deeper reserves within the soil when it doesn't. In the first two years of the plant's life, the visible parts remain small and stagnant while a resilient subsurface infrastructure of water access is under construction.

Plants respond to drought more visibly through diverse approaches to the shapes and textures of their leaves. The leaf is a critical site of exchange for plants, where it gains the CO_2 it needs to create energy for growth at the cost of releasing water through the opening of its stomata. To regulate this exchange, tropical plants have developed enormous leaves to maximize photosynthesis under a dense canopy, and dryland plants, constantly exposed to direct sunlight, have done the opposite. Plants with thick, small, evergreen leaves (sclerophyllous plants) reduce water loss by limiting leaf size and turning perpendicular or obliquely to the sun's rays. Species like *Quercus coccifera*, *Olea europaea*, and *Arbutus unedo* successfully employ this strategy to characterize the extensive exposed shrublands known as the Mediterranean *garrigue*

and *maquis*. Some plants, like many of the ones on Anafi, simply shed their leaves in the summer to stop losing water altogether, sacrificing growth until the autumn rains arrive. In addition to size, the leaves also play with texture to maximize water efficiency – tangles of tiny white leaf hairs reflect sunlight and reduce evaporation, while also often capturing residual condensation from daily dew drops.

The amazing thing about leaf adaptations is that, unlike roots, we can see them, which means that the plant's phenotypical responses to drought, cold, shade, and other climatic constraints become part of its aesthetic identity and can serve as fodder for design thinking. The tradeoff is not form vs. function, but a generative hybrid that comes from the living medium itself. It's possible to see the plant as "a meteorological instrument which integrates the various factors of climate and which, with experience, can be 'read' like a thermometer or a rain gauge."[2] Plants that are formed by extreme climates generate distinctive spatial qualities associated with dry and wet, shady and exposed, and this potential has been underutilized by the classical garden imagination of floral perennial borders. Landscape designers from non-Western climates, from Roberto Burle Marx to contemporary designers from the Mediterranean like Thomas Doxiadis, have lamented the hegemony of the Western European plant palette and celebrated the aesthetic specificity of tropical and Mediterranean plants and their potential for space-making.[3] But there is a problem with choosing plants from conditions outside the dominant supply chain of the nursery industry – where do you find them? For his cohousing project in Antiparos, Greece, Doxiadis initiated a plant nursery on the island to grow plants adapted to those conditions and tried to convince his clients that the plants they were using were part of a research project, and he is far from the only designer who has found it necessary to grow the plants for their own projects.[4] What is available locally and reliably is what largely defines the breadth and diversity (or lack thereof) of designed landscapes from private gardens to public parks.

Alternative nursery practices

The surprisingly consistent selection of plants in nurseries limits the range of plant choices for the particularities of a site, and more broadly it indicates a process of genetic streamlining. Driven by the profitability of standardization and global supply chains, the horticultural production industry has honed the plant-as-commodity: bigger flowers, faster development, and a ready-made final form.[5] This strips plants of their climatic intelligence, or what landscape architect Gilles Clément refers to as "*génie naturel*," by ignoring and often directly conflicting with plants' long-term responses to the milieux in which they evolved.[6] This pattern contributes to a rapid loss of complexity in the ecologies produced by industrialized societies like the US, where the same five annual bedding plants dominate horticultural consumption year after year.[7] Climatic constraints are a barrier to the expansion of markets for popular ornamental plants to reach profitable economies of scale, so an emerging driver for genetic breeding research is tolerance: the capacity to withstand increases in drought, heat, and salinity.[8] While breeding for greater tolerance might be a boon for large-scale horticultural producers amidst the major climatic shifts we are in, is this the method of adaptation that we want to pursue? A greater emphasis on sourcing and selecting plants globally that already have particular climatic adaptations would yield a much greater diversity of species, without being narrowed through a genetic bottleneck of the breeding industry.

The latter approach echoes a philosophy of "working with rather than against," a core tenet of Gilles Clément's concept of the planetary garden, in which humans are equally enmeshed in the multi-species milieu of a garden as their nonhuman counterparts and use the fewest resources to protect biodiversity in the garden as "a site of accelerated formulation."[9] In Clément's view, "the planetary garden has nothing to do with globalization. Although the scale of impact is the same, the planetary garden protects and develops diversity in all its forms, while globalization erases it in the name of market forces."[10] The planet has an existing genetic library that contains information about dealing with the challenges of climate change, and one that is perpetually shrinking because of habitat loss. Instead of choosing a few consumer favorites to reproduce at a large scale, we should focus our energy on understanding, preserving, and cultivating the range of species that hold that climatic knowledge.

Small growers and gardeners have long been sourcing plants and cultivating assemblages for specific climatic constraints – and often even creating those constraints in order to be able to play with a larger range of plants. In Beth Chatto's books *The Dry Garden* and *The Damp Garden* she describes the process of exaggerating conditions already found at the site of her home garden in Essex, England in order to create more diversity in those two extremes.[11] Her dry garden (which I visited in 2018 and looks more like Spain than how I imagined England) has never once been irrigated. In her book, Chatto states unequivocally, "I don't really hold with watering."[12] Peter Korn, a plant grower and gardener in Sweden, has taken extreme measures to trick the dryland plants he loves most into thinking that they live on well-drained sunbaked slopes rather than in the cold, moist conditions of Scandinavia (his garden beds are composed of 100% pure sand).[13] Sean Hogan, of Cistus Nursery on Sauvie Island in Portland, Oregon, has made climatic connections a core part of his operations, organizing the nursery by different types of Mediterranean-like climates around the globe, and traveling to collect seeds and reproduce those plants in Portland's analogous climate.

Coloniality obviously underlies a practice of traveling to collect seeds (often in the Global South) and propagating them in Western Europe and North America (where the aforementioned gardeners are located). The extractive relationships that enabled the vast collection of plants by European institutions from the

1 Olivier Filippi, "Mediterranean Landscapes as Inspiration for Planting Design," paper presented at the Beth Chatto Symposium, Colchester, England (August 30, 2018).

2 C.W. Thornthwaite, "An Approach toward a Rational Classification of Climate," *Geographical Review* 38, no. 1 (1948): 88.

3 Anita Berrizbeitia, *Roberto Burle Marx in Caracas: Parque Del Este 1956–1961* (University of Pennsylvania Press, 2005), 19; Arañazos en el Cielo, "Interview to Thomas Doxiadis by Elita Acosta" (July 16, 2019), http://debibliotecaycampo.blogspot.com/2019/07/interview-with-thomas-doxiadis.html.

4 Ibid.

5 Russell Hitchings, "Approaching Life in the London Garden Centre: Acquiring Entities and Providing Products," *Environment and Planning A: Economy and Space* 39, no. 2 (2007): 242–59.

6 Gilles Clément, *Jardins, paysage et génie naturel: Leçon inaugurale prononcée le jeudi 1er décembre 2011* (Collège de France, 2012); Emma Lewis et al., "Rewilding in the Garden: Are garden hybrid plants (cultivars) less resilient to the effects of hydrological extremes than their parent species? A case study with *Primula*," *Urban Ecosystems* 22 (2019): 841–54.

7 USDA, Floriculture Crops 2020 Summary (USDA National Agriculture Statistics Service, 2021).

8 Nehanjali Parmar, et al., "Genetic Engineering Strategies for Biotic and Abiotic Stress Tolerance and Quality Enhancement in Horticultural Crops: A comprehensive review," *3 Biotech* 7, no. 4 (2017): 239.

9 Gilles Clément, "In practice: Gilles Clément on the planetary garden," *The Architectural Review* (February 16, 2021).

10 Ibid.

11 Beth Chatto, *The Dry Garden* (Weidenfeld & Nicolson, 1978).

12 Ibid, 19.

13 Peter Korn, *Peter Korn's Garden: Giving Plants What They Want* (Landvetter, 2013).

14 Etienne Benson, *Surroundings: A History of Environments and Environmentalisms* (University of Chicago Press, 2020), 19-24.

15 Gilles Clément, "In Praise of Vagabonds," trans. Jonathan Skinner, *Qui Parle* 19, no. 2 (2011): 275.

16 Ibid., 275–97.

Next: The global map overlaps the variables of aridity index and minimum temperature, using color hue to communicate minimum temperature from the warmest (red) to the coldest (black), and color value to communicate aridity (lighter is more arid, darker is more humid).

end of the 18th century onward is well-explored within the discipline and beyond.[14] Then as now, the movement of plants across the globe by small nursery growers is made possible by the knowledge of local people, and there is an undeniably unequal exchange inherent in that dynamic. But while the relationship between growers and the sites they visit should emphasize reciprocity, the practice is an imperfect alternative to the industrialization of plants by transnational horticultural companies. It also offers an alternative to its reaction: a narrow focus on native plants. It's incredibly important for growers to propagate and stock plants native to their ecosystems, a practice that contributes to stewardship of localized genetic libraries, preserves existing mutualistic relationships with other living things, and enables designed landscapes that are more pest-resistant and need fewer external inputs like fertilizers, soil additives, and other products that rely on fossil fuel consumption. However, a belief in the primacy of native plants glosses over a key factor that defines the composition of a landscape: movement. "Plants travel, especially grasses."[15] As a part of Clément's call for a planetary gardener to cultivate plants adapted to site, he recognizes that the garden is a transient container for a landscape dynamics beyond the gardener's control.[16] Carex Vivers, a small plant nursery near the Pyrenees in Northern Catalonia, describes "unpredictability and change as drivers of biodiversity, but also of knowledge."[17] The movement of landscape agents occurs simultaneously at different scales of space and time. On the one hand, we have contemporary urban conditions, which engender the inevitable movement of seeds and plants through the movement of other materials. On the other, we have a rapidly changing climate in which the natural speed of plant migration is insufficient for the survival of species.[18]

At his agroforestry test gardens and nurseries in South Devon, England, Martin Crawford has been testing the viability of non-native oak species in response to the sharp decline of native oak species in England.[19] In discussing the replacement of *Quercus petraea*, a common oak in the UK, with *Quercus pubescens*, dominant in southern Europe, Crawford comments, "We can't just continue to plant native oaks, sit around and hope for the best…that's not a tenable position."[20] By taking a planetary view, it is possible to combat the globalizing effects of the ornamental plant trade in favor of a highly specific relationship to plants that draws on the generative practice of testing and observation, and a recognition that plants move, and that we move them. This line of thinking corresponds to themes in postcolonial discourse that recognize the inevitable globality of cultural production while resisting universality, or globalization. Theorist and writer Édouard Glissant has emphasized what he terms the poetics of relation: "the account is open, everything communicates with the rest of the universe, and our imaginary must lead us to this poetics of globalness which is the only one possible…What we need to think about is the realized totality of all the world's cultures, not an overreaching universality."[21] He describes a "planetary consciousness of a world of inextricable

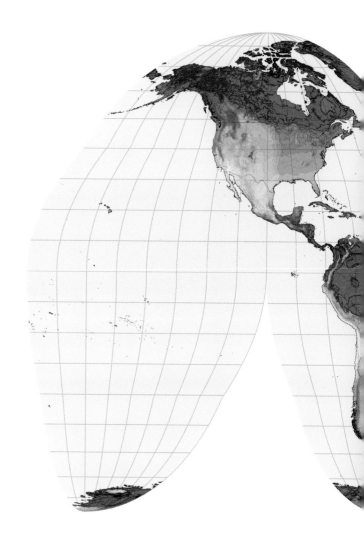

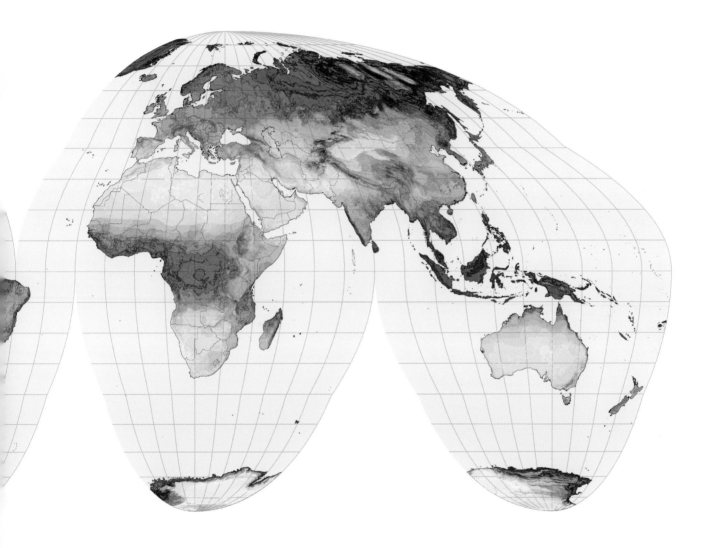

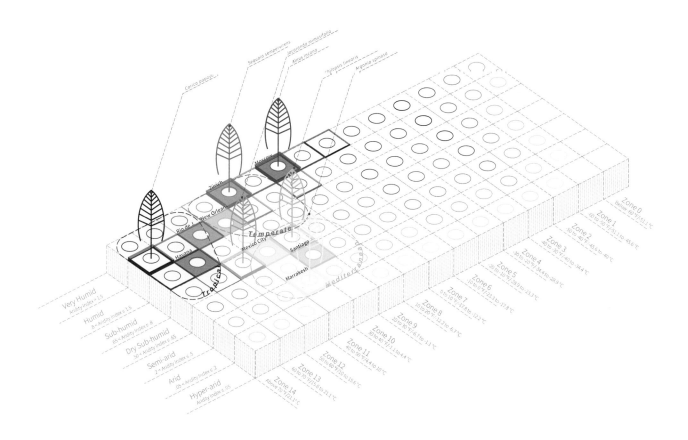

relations in which it is necessary for all of us to change our ways of conceiving, living, and reacting."[22] Taking this planetary consciousness as a point of departure, what are the tools and methods for designing a garden of relation that cultivates the local while keeping in mind the global?

Drawing Climate

Olivier Filippi's nursery, outside of Montpellier in the French Mediterranean, specializes in plants that endure intense summer droughts and thrive in highly exposed conditions like those of Anafi that I described at the beginning of this article. When I visited the nursery in September of 2022, Filippi told me that he has learned by traveling to observe plants and how they survive in arid conditions across the globe, often collecting their seeds to try to reproduce in southern France. A core part of Filippi's extensive research in drought-tolerant plants is the use of two tools: the Gaussen diagram and the drought code, the latter of which he developed and published in his first book, *The Dry Gardening Handbook*.[23] These tools seek to standardize a method for approximating a plant's resistance to drought, which is conspicuously missing from the universe of gardening guides and helpful how-tos. Methods of comparing plants through observation of their climatic behavior begin to put a wide range of plants in distinct parts of the globe into conversation with each other. How do the species that have emerged within the specific milieux of the Cycladic islands relate to those of L'étang de Thau?

The USDA Plant Hardiness Zone map is a standard feature of gardening manuals, nursery tags, and botanical websites across the US, and similar maps exist for Canada and Europe.[24] The zones help people to understand the capacity of a plant to survive the coldest temperatures of a place, one of its main climatic constraints to growth. But it's not the only one. The map is missing a key factor for plant survival and morphology: water. If identifying a plant's zone is our main tool for decision-making about plant suitability, it's not sufficient, particularly in the increasingly unstable water situation presented by a changing climate. In the USDA zone map, where I'm from in Eastern Tennessee is the same classification as the panhandle of Oklahoma, and those who have visited these places (or

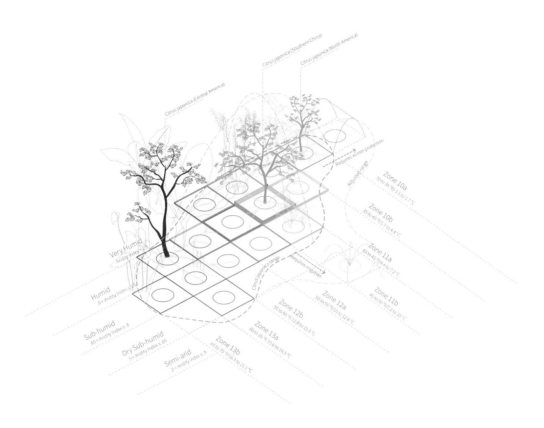

have looked on a satellite photograph) can see that the short grass prairie of the southern plains is not the humid subtropical deciduous forests of the Southeast.

The map, using a method developed by the Arnold Arboretum, was created largely with the intent of guiding agricultural practices and crop selection. Extensive extraction of freshwater resources supported the Green Revolution and continued agricultural expansion throughout the drylands in the middle and western part of the US in the 20th century. It is possible, perhaps, that water was simply not considered a limitation for growing plants in the heyday of seemingly endless resources. In fact, climatic mapping within the botanical sciences has always tended to have a temperature bias. In the 17th century, climatic measurements usually referred only to temperature, and botanical geography maintained a principle of "latitudinal homogeneity."[25] This bias is perhaps unsurprising, given that modern botanical sciences emerged in Sweden and other parts of Northern and Western Europe with plenty of water, but cold winters. However, as water becomes increasingly precious in much of the US and globally, it merits a reconsideration of our main guide, which is why Filippi's research on testing and measuring a plant's capacity for drought has become a beacon for gardeners and growers.

In the 1940s climatologist Henri Gaussen and F. Bagnouls developed a diagram that visualizes the availability of water for plants over the course of the year, depending on temperature.[26] Gaussen defined drought as any day when the amount of precipitation in millimeters is less than or equal to twice the temperature in Celsius (averaged over 30 years).[27] While simplified, this definition can be quick and helpful, as it is easy to get temperature and precipitation data for a site. What Gaussen keyed in on is that not only are the amounts of temperature and precipitation important when it comes to plants, but also the relationship between them. Places that are hot in summer, but have abundant rainfall, have a predictably different capacity than those that are both hot and dry during the summer months. And for plants, the amount of time that they can endure a drought condition is an important determinant in whether they can survive in a given place. The Gaussen index, therefore, is a marker of the number of days (on average) that a location has a drought condition.[28] However, Gaussen's work was concentrated on the Mediterranean and didn't cover more humid zones that don't experience drought in a typical 30-year average.

Another measurement that approximates the hydric deficit for plants and can be used more broadly is potential evapotranspiration (PET). Recognizing that evapotranspiration played a critical role in climate but was very difficult to measure, PET was developed as an equation by climate scientist Charles Thornthwaite in 1948 who described it as the amount of evapotranspiration that would occur if there were sufficient water sources available. Like Gaussen, Thornthwaite recognized the importance of the relationship between precipitation and temperature, "We cannot tell whether a climate is moist or dry by knowing precipitation alone. We must know whether precipitation is greater or less than the water needed for evaporation and transpiration. Precipitation and evapotranspiration are equally important climatic factors."[29] It is possible to arrive at a solution for PET using one of a few equations that involve average day length, number of days in a month, average temperatures, and a variable for heat index. But more importantly, these calculations have been done by scientists for most of the globe, and that data is publicly available.[30] By dividing the actual average precipitation by the potential evaporation (PET), climatologists have developed the aridity index: a number between 0–6.5 that determines the degree of aridity (or humidity, if you're the glass-half-full type). Luckily, the aridity index is also largely available as GIS data at a fairly high resolution (30 arc-second).[31]

With the available data, perhaps we can make an adjustment to our zones map that attempts to recapture the factor of moisture. As a part of an ongoing research project at the Chair of Being Alive at ETH Zurich, we have developed a version of the plant hardiness map for the globe that overlaps lowest minimum temperature (aligning with the USDA Zone increments) with average aridity index. Clearly there are other climate maps and classifications, made by climatologists and not landscape architects, that integrate more factors and do a better job approximating the complexity of climate (like the Koppen-Geiger classification and its updates). Realistically, most climate research since the development of computing doesn't use maps at all, but sophisticated models that can simulate weather patterns, wind, vegetation types, land use, and myriad other contributing factors. Climate itself is only one of the long list of factors that combine to form the conditions for plant growth, like hours of daylight, soil type and texture, land use, and supportive relationships with other living things. But there's a value in drawing what we do know, and how we know it. We can leave the climate models to the climatologists, while seeking to connect research about plants to the main climatic factors that determine their viability. There is value to a method that does this in an accessible way and abstracts the infinitely complex reality that defines the life of a plant in an effort to better understand it. Incorporating drawing and diagramming methods that allow us to integrate climatic research on plants (climate mapping, the Gaussen diagram, etc.) opens potentials in the process of designing with plants.

Building on the method of overlapping the minimum temperature and the aridity index in the map, we have also been developing a drawing tool that abstracts these intersections into an axonometric, which can be used to visualize the climatic difference of where plants come from and where they are planted. Using the diagram at the level of an individual plant, the axes communicate the extent of a plant's range temperature and aridity, which has the potential to integrate research from plant growers in different climates to verify a plant's capacity to withstand drought or cold temperatures (like Filippi's drought code). We are currently working with the climate axonometric to integrate data on the soil water-holding capacity and the PET of a given site, which adds another dimension to the research potential of the drawing.[32] When used in combination with the Gaussen diagram described above (which communicates the length of drought), the climate axon is a drawing tool that allows designers to better understand the climatic logics driving a plant's behavior and form – but also opens up possibilities for drawing the links between places to their future climates, and plants to future habitats.

Recent research from the Crowther Lab at ETH Zürich has paired certain cities in terms of similarities between the current climates and the ones they will experience by 2050. According to their projections, London will resemble Melbourne, Zürich will resemble Milan, and Milan will resemble Dallas. Portland, Oregon will feel like San Antonio, Texas, and New York City like Virginia Beach.[33] A rigorous practice of drawing climate can help us understand and design those relations among places, their plants, their gardening practices, and their future climates. What is shared by the growers and gardeners mentioned in this article is a method of growing plants according to the climatic constraints of their origin and celebrating the potential of those constraints (loving plants that love it dry, loving plants that love it cold). As designers, we must resist the homogeneity of a horticultural industry that seeks to profit from engineering plants as commodities, while cultivating a capacity for observation, relational thinking, and the sharing of knowledge, practices, and, perhaps, seeds.

17 Carex Vivers, "El Retorno a la Naturaleza" http://www.carex.cat/es/vivers-carex/empresa.aspx (accessed December 20, 2022).

18 Emma Marris, "A Scientific Argument for Intervening in Nature," *Scientific American* (October 14, 2011).

19 Alice Broome et al., "Ecological Implications of Oak Decline in Great Britain," *Forest Research Note* 40 (May 2021).

20 Martin Crawford, in-person conversation (September 11, 2022).

21 Édouard Glissant, "The Poetics of the World: Global Thinking and Unforeseeable Events," trans Kate Cooper Leupin, *The Glissant Translation Project*, https://sites01.lsu.edu/wp/theglissanttranslationproject/2017/10/20/the-poetics-of-the-world-global-thinking-and-unforeseeable-events/.

22 Chris Bongie, "Édouard Glissant: Dealing in Globality," in Charles Forsdick & David Murphy (eds), *Postcolonial Thought in the French Speaking World* (Liverpool University Press, 2009), 93.

23 Olivier Filippi, *The Dry Gardening Handbook: Plants and Practices for a Changing Climate* (Filbert Press, 2019).

24 USDA Plant Hardiness Zone Map, https://planthardiness.ars.usda.gov/; Natural Resources Canada, http://planthardiness.gc.ca/?m=1 (both accessed December 20, 2022).

25 Marie-Noëlle Bourguet, "Measurable Difference: Botany, Climate, and the Gardener's Thermometer in Eighteenth-Century France," in Londa Schiebinger & Claudia Swan (eds), *Colonial Botany: Science Commerce, and Politics in the Early Modern World* (University of Pennsylvania Press, 2005), 274.

26 Filippi, *The Dry Gardening Handbook*, 41.

27 Henri Gaussen, "Bioclimatic Map of the Mediterranean Zone," *Arid Zones Research* XXI (UNESCO-FAO, 1963).

28 Ibid.

29 C.W. Thornthwaite, "An Approach toward a Rational Classification of Climate," 55.

30 WorldClim – Global Climate Data, http://www.worldclim.com/version2 (accessed December 20, 2022).

31 Ibid.

32 Zhao Ma & Teresa Gali-Izard, "BeingAliveLanguage: Visualizing Soil Information from a Design Perspective to Enhance Multidisciplinary Communication," *Ecological Informatics* (forthcoming).

33 Jean-Francois Bastin, et al., "Understanding climate change from a global analysis of city analogues," PLoS ONE 14, no. 10 (July 10, 2019), https://doi.org/10.1371/journal.pone.0217592; Crowther Lab, "Current vs. future cities," https://crowtherlab.pageflow.io/cities-of-the-future-visualizing-climate-change-to-inspire-action#213121.

PLANTS ON THE MOVE

JANET MARINELLI

> A SEPAL, petal, and a thorn
>   Upon a common summer's morn,
> A flask of dew, a bee or two,
> A breeze
> A caper in the trees,–
>   And I'm a rose![1]

Janet Marinelli is an independent science writer who covers the effects of climate change on plants and animals. She is part of the team that won a 2020 National Magazine Award for *Audubon*, covering the creation of landscapes that reduce carbon emissions. She was a director of scientific and popular publishing and digital media at Brooklyn Botanic Garden for 16 years.

+ ECOLOGY, CONSERVATION

During her lifetime, Emily Dickinson (1830–1886) was better known for her exquisite garden than for lyric poems that revealed a passionate love of nature. On rare occasions, the townsfolk of Amherst, Massachusetts, would catch a glimpse of the mysterious recluse, a ghostly figure dressed in white, leaning over to tend her flowers by flickering lantern light. Dickinson's poetic voice grew from her insights as a gardener, distinguishing her from fellow 19th-century American writers whose thinking became the bedrock of conservation. While Henry David Thoreau famously declared wild places to be "the preservation of the world," Dickinson was waxing ecstatic about the power and beauty of a garden rose.

Among the plants that survive on the family property where Dickinson confined herself for much of her adult life are picturesque old trees called umbrella magnolias (*Magnolia tripetala*) – so named because their leaves, which can reach two feet long, radiate out from the ends of branches like the spokes of an umbrella. The trees, believed to have been planted in the mid-1800s by Emily's brother Austin, have jumped the garden gate and established wild populations not far from the poet's home. This new location is a few hundred kilometers north of the tree's historic native range, centered in sheltered woods and ravines of the Appalachian Mountains, and is the first evidence that native plant horticulture in the United States "is giving some species a head-start on climate change," in the words of Smith College biologist Jesse Bellemare.

About a decade ago, Bellemare noticed umbrella magnolias peeking out from roadside vegetation in western Massachusetts. He observed that most of the naturalized plants were in close proximity to the horticultural specimens. Intrigued, he set out to determine the age of the garden escapees. Although some of the landscaped trees, like those on the Dickinson homestead, were planted more than a century and a half ago, core samples from a number of the largest naturalized individuals revealed that the species started escaping profusely only in the last 30 years. As Bellemare and coauthor Claudia Deeg pointed out in a 2015 paper in the botanical journal *Rhodora*, this is also when the climate began warming quickly in the region.[2] The following year, in a presentation at the Ecological Society of America conference, Bellemare, Deeg, and Brown University biologist Dov Sax concluded, "It is unlikely that natural dispersal from the South would have allowed *Magnolia tripetala* to reach this region anytime soon."[3]

Umbrella magnolia has become an emblem of the mass exodus of plants and animals that is transforming the distribution of the world's biodiversity in the face of global warming. It has also achieved notoriety as one of the small but growing number of species that have been dispersed beyond their former range limits by native plant horticulture. Ironically, the denizen of the Dickinson homestead is challenging basic precepts of conservation practice, such as what is the definition of "native"? Are climate refugees that hitchhike north via horticulture less worthy of protection than plants that arrive on their own? Do they pose a threat to existing native species? Should landscape designers help native plants stranded in inhospitable habitat migrate to more suitable climes?

Central to conservation thinking has been the idea that native species, traditionally defined as those that have arisen through evolution in a region, are worthy of protection, while non-natives, which have been transported by humans, likely are not. In recent decades, this Manichean view has been buttressed by numerous studies documenting how invasive non-native plants and animals have hijacked the habitat of existing native species. Over the past 10 or so years, however, the effect of non-native species on native biodiversity has become one of the most contentious issues in ecology. In a 2011 commentary in the journal *Nature*, Mark Davis of Macalester College and 18 fellow ecologists threw down the gauntlet, asserting that the native/non-native dichotomy needs to go.[4]

The scientists traced the concept of "nativeness" back to English botanist John Henslow in 1835. By the late 1840s, botanists had widely adopted the terms "native" and "alien" to distinguish the "true" British flora from arrivistes from other lands. In the 1990s, the new field of invasion biology cemented the perception that introduced species pose, in the words of the *Nature* coauthors, "an apocalyptic threat to biodiversity."[5] But only a small fraction of non-native species has wreaked havoc in their new habitats, they pointed out. We must embrace the reality of "novel ecosystems" consisting of native and non-native species, they concluded, and abandon the quixotic goal of eradicating them or drastically reducing their numbers.[6]

A small but influential number of scientists and environmental journalists adopted this view, prompting an outcry from other biologists denouncing it, most famously University of Auckland conservation biologist James C. Russell, who declared that abandoning conservation's native/non-native dichotomy is tantamount to "denialism," given that invasive non-native species "now rank as one of the major challenges to biodiversity conservation of our time."[7]

Early climate refugees like umbrella magnolia have complicated the debate. In a 2019 paper in *BioScience*, University of Vienna professor Franz Essl and other invasion biologists proposed a new term, "neonative," for species that have moved at least 100 km or a few hundred meters in altitude beyond their historic ranges in response to environmental changes.[8] However, they reserve this term only for species that are expanding their ranges "naturally," leaving those like umbrella magnolia–which have had direct human assistance–in a kind of existential limbo.

While some scientists tie themselves in knots trying to determine what is native or neonative and what is not, Bellemare may have found a way out of the dilemma. A member of the regional advisory committee guiding the update of the Native Plant Trust's *Flora Conservanda*, which identifies plants in New England that are rare and worthy of conserving, he believes that terms such as native and even neonative are too loaded. He prefers the term "range-shifting" because it simply describes the phenomenon without the divisiveness and value judgements around calling something exotic.[9] Bellemare notes that the standard definition of native was based on a view of nature as unchanging, and of what constitutes "native" as absolute and enduring. But conservation thinking needs to evolve, he says, as the boundaries of entire biogeographic regions shift with changing conditions – including the eastern deciduous forest, which historically has stretched over 26 states from Florida and Texas north to New England and the upper Midwest and into southern Canada.

According to paleoecologists, plant and animal communities have always been on the move. As the climate cooled and glaciers expanded during the Pleistocene era, for example, forest plants of the Northeast survived by migrating to so-called refugia in the South; when the climate warmed and the most recent ice sheet receded, some of these species were able to recolonize the northern habitats they had lost. When viewed in the context of this larger time scale, it is not difficult to conclude that New England and other northern regions should harbor southern species fleeing north from global warming. In fact, many of these climate refugees may simply be ancient natives returning home. And whether plants from, say, the Southeast and mid-Atlantic arrived via horticulture or on their own, the risk that they will cause ecological havoc is likely lessened by the fact that they share a biogeographic history, including a suite of natural enemies, with those in New England.

But the real issue, Bellemare notes, is not that plants are rushing to colonize new areas as global warming intensifies, but rather that they "are not moving as much as we might predict, based on the amount of climate change we have seen so far."[10] For years, biologists have been warning that while numerous species have migrated successfully in response to past episodes of climate change, the rapidity of current, human-caused global warming, combined with the widespread fragmentation of the modern landscape by human development, agriculture, and other barriers to natural dispersal, is likely to trigger a major extinction crisis in coming decades. Biologists believe that endemic plants—those that are found in small, narrowly restricted ranges and nowhere else in the world—are among the plants at greatest risk as their preferred climate jumps far beyond their ability to disperse.

Bellemare, along with University of Minnesota biologists Stephanie Erlandson and David Moeller, analyzed the likely impact of climate change on one of the most celebrated clusters of endemic plants – herbaceous wildflowers of the Southern Appalachian Mountains, a biodiversity hotspot with the highest concentration of plant diversity and endemism in eastern North America.[11] From bleeding hearts to trilliums, they produce a flamboyant explosion of blooms that carpet the forest floor in spring. These endemic wildflowers have been unable to disperse northward in the 15,000 years since the last ice sheet began receding. While they would probably be well positioned to survive the climatic cooling of another ice age, the breakneck speed of current warming seems to place many of these plants, in Bellemare's words, "on the wrong side of climate history." He believes that such plants are prime candidates for assisted migration—also known as assisted colonization or managed relocation—the emerging conservation strategy in which scientists move species unable to disperse to new habitats fast enough on their own.

Meanwhile, we are already engaged in a wildly risky assisted migration experiment with the non-native ornamentals being sold through the nursery trade, says Bethany Bradley, a University of Massachusetts biologist who specializes in invasion ecology. She goes well beyond most biologists in asserting that landscape professionals, gardeners, and property owners should be assisting the migration of native plants. Several years ago, when Bradley moved back to Massachusetts and started planning her garden, her mother sent her an old backyard planting guide published in the 1970s. Bradley "was chagrined to discover that the guide was a 'who's who' of Northeast invasive plants," from Oriental bittersweet and Japanese barberry to autumn olive.[12] As she points out, gardeners are significant vectors of introduction and dispersal of invasive plants – we're so good at dispersing them that they have easily overcome the distances and natural and manmade barriers that have restricted native species. What's more, new plants are constantly arriving via the nursery trade, and most are from countries that are warmer than most regions in the United States and therefore preadapted to warmer temperatures, which gives them an enormous advantage over natives. Bradley notes that the climate of her home state of Massachusetts is more like New Jersey was 30 years ago. Studies predict that if decisive action to control carbon emissions is not taken, by the year 2100 the Massachusetts climate will be more like today's South Carolina – a distance of 1,279 km. "If we want species to survive extinction," she says, "then we need to help them move."[13]

As an invasion biologist, Bradley is well aware that the worst-case scenario is that a transported species becomes invasive. The most cited example of a range-shifting native gone rogue is the black locust (*Robinia pseudoacacia*), a medium-size deciduous tree in the legume family that produces pendulous clusters of fragrant, white, pea-like blossoms in spring. Its historic native range is thought to be two separate areas: the Appalachian Mountains from northern Georgia to Pennsylvania, and the Ozarks of eastern Oklahoma, Arkansas, and Missouri. The tree has been widely planted, is expanding its range due to climate change, is listed as invasive in Connecticut, Wisconsin, and Michigan, and is prohibited in Massachusetts.

In Bradley's words, moving natives "is not risk free, but it's far less risky than the status quo of importing and introducing new species from other continents."[14] In fact, studies have found that native "invaders" are relatively rare in plants. In a 2012 study published in the journal *Ecology*, a group of researchers led by University of Tennessee, Knoxville ecologist Daniel Simberloff found that at least 12% of non-native plants in the United States were invasive, compared to less than 0.3% of natives. In other words, the non-native plants proved 40 times more likely to become invasive.[15] Because there is even a small chance of a native plant becoming invasive, Bradley says we need to weigh the risks before moving them. Bradley is a

coauthor of a 2020 paper published in *Nature Climate Change* proposing that the same basic tools that invasion biologists use to assess the risks of nonnative plants, such as the Environmental Impact Classification for Alien Taxa framework (EICAT), should be used to evaluate native species.[16]

In "Nuisance Neonatives: Guidelines for Assessing Range-Shifting Species," Bradley and colleagues in the Northeast Regional Invasive Species & Climate Change Management Network provide a list of risky traits that can help predict which plants are likely to become problematic.[17] Among the red flags: Does the species produce copious amounts of seed that are easily dispersed? Is it common in its native range? Has it been identified as an invasive species elsewhere? Does it carry disease pathogens or pests it can spread to its new habitats? Is it a so-called ecosystem engineer, likely to fundamentally change these habitats in ways that make them less suitable for long-time native residents? Black locust, for example, is a nitrogen-fixer, altering its new environments by enriching the soil and potentially making it less suitable for existing natives and more favorable for invasive species. If a plant has one or more of these traits, it may not be a good candidate for assisted migration. Bradley makes a point of adding that doing nothing is also risky. "We're on the cusp of a mass extinction. If we do nothing, that mass extinction is going to be worse."[18]

How far should we move plants? Distance may not be as important as choosing a species from the same biogeographic region. As a resident of Massachusetts, Bradley says she feels comfortable with choosing any eastern forest plant from farther south that can tolerate today's temperatures.[19] "Planting local-ish," as she calls this, is "a win-win." It not only reduces the likelihood that we will inadvertently introduce an invasive species but helps natives keep up with climate change, while supporting birds and other wildlife.[20] Taking the long view, Bradley points out that as human influence on remaining natural landscapes grows, the gardens of today increasingly are seeding the ecosystems of tomorrow. Right now, with the continuing introduction of potentially invasive non-native species, she says, "it looks like the ecosystems of the future will be pretty weedy. But we have an opportunity to change that."[21]

While scientists grapple with the implications of range-shifting species, there is poetic justice that a plant from the Dickinson homestead has sparked the discussion. Although the vision of enduring wilderness championed by Thoreau and John Muir came to dominate conservation thinking, Emily Dickinson, who perceived the beauty and destructive capacity of nature in her own garden, may be the more appropriate literary icon for an age of climate disruption.

---

[1] Bartleby.com, "Emily Dickinson (1830–86). Complete Poems. 1924," https://www.bartleby.com/113/2093.html.

[2] Jesse Bellemare & Claudia Deeg, "Horticultural Escape and Naturalization of *Magnolia tripetala* in Western Massachusetts: Biogeographic Context and Possible Relationship to Recent Climate Change," *Rhodora* 117, no. 971 (2015): 371–83.

[3] The Conference Exchange, "Recent Synchronous Horticultural Escape and Naturalization of *Magnolia tripetala* North of Its Native Range Will Give Tree Species a 'Head Start' on Climate Change," https://eco.confex.com/eco/2015/webprogram/Paper53843.html (accessed February 20, 2023).

[4] Mark A. Davis, et al., "Don't Judge Species on Their Origins," *Nature* 474 (2011): 153–54.

[5] Ibid., 153.

[6] Ibid., 154.

[7] James C. Russell & Tim M. Blackburn, "The Rise of Invasive Species Denialism," *Trends in Ecology & Evolution* 32, no. 1 (2017): 3–6.

[8] Franz Essl, et al., "A Conceptual Framework for Range-Expanding Species that Track Human-Induced Environmental Change," *BioScience* 69, no. 11 (2019): 908–19.

[9] Interview with author, February 22, 2023.

[10] Ibid.

[11] Stephanie K. Erlandson, Jesse Bellemare & David A. Moeller, "Limited Range-Filling Among Endemic Forest Herbs of Eastern North America and Its Implications for Conservation with Climate Change," *Frontiers in Ecology and Evolution* 9 (2021), https://doi.org/10.3389/fevo.2021.751728.

[12] Bethany Bradley, "Proactive Weeding of Our Gardens of the Future," *Ecological Landscape Alliance* (May 15, 2019).

[13] Bethany Bradley, "Ecological Gardening with Climate Change to Prevent Future Invasions and Assist Native Migration," *Ecological Landscape Alliance* Webinar (October 11, 2019).

[14] Bradley, "Proactive Weeding of Our Gardens of the Future."

[15] Daniel Simberloff, et al., "The Natives Are Restless, But Not Often and Mostly When Disturbed," *Ecology* 93, no. 3 (2012): 598–607.

[16] Piper D. Wallingford, et al., "Adjusting the Lens of Invasion Biology to Focus on the Impacts of Climate-Driven Range Shifts," *Nature Climate Change*, 10 (2020): 398–405.

[17] Scholarworks @ UMass Amherst, "Nuisance Natives: Guidelines for Assessing Range-Shifting Species," https://scholarworks.umass.edu/eco_ed_materials/9/ (accessed February 20, 2023).

[18] Bradley, "Ecological Gardening with Climate Change to Prevent Future Invasions and Assist Native Migration."

[19] Email correspondence with author, February 17, 2023.

[20] Bradley, "Proactive Weeding of Our Gardens of the Future."

[21] Bradley, "Ecological Gardening with Climate Change to Prevent Future Invasions and Assist Native Migration."

IRUS BRAVERMAN
GREEN GOLD
THE AKKOUB'S SETTLER ECOLOGIES

Irus Braverman is professor of law and adjunct professor of geography at the State University of New York at Buffalo. She is the author of several monographs, including *Planted Flags: Trees, Land, and Law in Israel/Palestine* (2009), *Zooland: The Institution of Captivity* (2012), and *Coral Whisperers: Scientists on the Brink* (2018). Her latest monograph, *Settling Nature: The Conservation Regime in Palestine-Israel*, was published with the University of Minnesota Press in 2023.

+ LAW, POLITICS

The akkoub (*Gundelia tournefortii*) is a thistle-like plant so precious to Palestinians that it is often referred to as "green gold." Considered a rare delicacy, the list of health benefits associated with this edible plant only adds to its desirability. Palestinian foragers risk their lives to put their (gloved) hands on this thorny plant, venturing into fields dotted with undetonated land mines in the northern region of the Jawlan-Golan to collect a fresh batch during its short season in the wild. The risks—as well as the mystique—surrounding the akkoub have only intensified since the State of Israel designated this plant as protected under the Nature and Parks Protection Act.

The story of the akkoub illustrates the three tenets of "settler ecologies": the regime of environmental protections enacted by the settler state (or by neocolonial modalities of governance on regional and global scales) that furthers the domination of the natural landscape and the dispossession of local and Indigenous communities.[1] The first and fundamental tenet of settler ecologies is the juxtaposed mindset it seeks to advance: the entire ecological system is seen through a binary prism, recruiting living beings to what is portrayed as an all-encompassing ecological war. The type of nature that is valued under this mode of thinking is a "wilderness" ideal of nature that is juxtaposed with culture and fixated on a particular past. This past is associated with the settler community, which in turn aims to restore it through ecological measures. Such an imaginary past landscape is finally juxtaposed with the present landscape, which is depicted as degraded and deteriorated because of the natives' reckless and even criminal behavior.[2]

The power of juxtapositions was especially visible in my legal ethnographic research on the conservation regime in Palestine-Israel.[3] However, the native-settler dialectic of the colonial nation-state extended here beyond the nature-culture and the past-present divides. I traced, for example, how the [Bedouin/local/domestic] camel is juxtaposed with the [Zionist/settler/wild and reintroduced] Asiatic wild ass, and the black [Palestinian] goat with [Jewish/European] pine seedlings. In an earlier project, I explored the warring tree landscapes in this region, documenting the pitting of the [forest] pine tree against the [cultivated] olive. While the former was deemed protectable for its affinity with the settler state's project of afforesting the desolate landscape, the latter was deemed killable (or uprootable) for its affiliation

with the agrarian Palestinian community and its purportedly unwild nature.[4] Such animosities run wild; leaning on each other, they reinforce, naturalize, and thus legitimize the power and the seeming inevitability of the juxtaposed mindset so characteristic of settler ecologies. As Amitav Ghosh points out, settler colonialism has been fought "primarily not with guns and weapons but by means of broader environmental change… Indigenous peoples faced a state of permanent…war that involved many kinds of other-than-human beings and entities: pathogens, rivers, forests, plants, and animals all played a part in the struggle."[5] Additional juxtapositions, such as those between renewal and demise and between hope and despair, are quite typical of modern conservation approaches, not only in Palestine-Israel but also around the globe.[6]

Second, settler ecologies are means of green dispossession, performed by both genocidal elimination and through the accumulation of natural capital.[7] Colonialism and capitalism thus work hand-in-hand through conservation to inflict violence on racialized populations, both human and nonhuman. While the appropriation of territory and land are central to the settler colonial regime, they do not exist on their own: they are accompanied and even preceded by imaginaries that lend them meaning and support. Dispossession and other forms of takeover and erasure by various ecological means, such as the criminalization of the natives by the settler state as well as its advancement of an orientalist approach toward them, are the second tenet of settler ecologies.

Third and finally, settler ecologies operate through environmental law, its rigid definitions and categories of protection both enabling and regulating their continued governance through the ecological state. The pronounced role of law is central to the operation of settler ecologies, which are typically enforced by extensive nature administrations. While settler ecologies include several additional tenets, such as *terra nullius* and ecological exceptionalism, the three tenets I identified here are the primary components of this project.

Thorny Ecologies
In 2005, the State of Israel protected the akkoub under the 1998 Nature and Parks Protection Act. Just as with the state's protection in 1977 of the local herb za'atar (*Majorana syriaca*)—the central ingredient in the salty blend that has come to represent the Palestinian cuisine worldwide—Israel explained the need to protect the akkoub by alerting to its decline in the region. And just like with the za'atar, the Zionist state's protection of the akkoub, too, has been viewed by the local Palestinian communities as a political and militant affront. This legal protection has therefore alienated many Palestinians, deeming Israel's environmental efforts an inherent part of its broader "occupier's law."[8] Palestinian acts of steadfast resistance (*sumud* in Arabic) to the akkoub's protection have

since then emerged, as thorny and resilient as the plant they have aligned themselves with. In such instances, the state has only doubled down on its green prohibitions, flexing its muscles even when the foragers were eight-year-old children gathering food for their family dinner. Instead of dinner, in one instance four children who went foraging for akkoub in Jabal al-Khalil (the southern Hebron hills) brought home a high fine, after having spent their evening at the Israeli police station. In 2014, a 14-year-old Palestinian boy was shot and killed by Israeli soldiers after he passed through a breach in the Separation Wall near the Palestinian village of Walaje near Bethlehem to forage akkoub on farmland owned by his family.[9]

Unlike Israel's widespread and successful campaign to protect wildflowers, which mostly targeted its Jewish population, the enforcement of the za'atar and akkoub protections by the Israel Nature and Protection Authority (INPA)—the governmental body that administers nature in Palestine-Israel—has been harsh and has exclusively targeted Palestinians. Violations have resulted in up to three years in jail and fines have easily reached thousands of Israeli shekels—a hefty amount for *fellahin* and rural communities.[10] Often, an order to stay "several hundred meters away" from the protected plants is imposed in addition to the penalties, and is executed not only in nature reserves but also in lands owned privately by Palestinians.[11] The Palestinian organization Adalah documented that "in the years 2016–2018, 26 indictments were submitted and 151 fine notices were issued for offenses related to these plants."[12] At the same time, Jewish Israelis were neither indicted nor fined using this law. INPA's regional biologist in the north explained: "This traditional picking of the akkoub is [only] performed by the Druze and Arab sector. The Jewish sector doesn't do it. They wouldn't even know how to eat it."[13] What this official failed to mention, however, is that no criminal charges were ever issued against Jews for picking protected wildflowers either.[14] He also neglected to say that the most serious damage to the akkoub and the za'atar has been in the hands of Jewish Israeli developers, not Palestinian foragers. As an Israeli expert on agriculture put it: "No one talks about the fact that we [Jews] destroy much more za'atar than the Arabs pick. Do you know how many great za'atar populations were uprooted by [our] bulldozers?"[15]

As with many wild plants and animals in Palestine-Israel, in the case of the akkoub and the za'atar, too, the Zionist settler project has been investing considerable efforts into excavating the ancient origin of these species. Utilizing linguistic studies alongside biblical geography and archeology, the settler state has been tracing contemporary wild creatures back through time in an attempt to prove its superior autochthonous connection to this place. Through their association with ancient Jewish texts from the Mishna and the Talmud, the akkoub and za'atar recently transformed into Jewish plants. It has therefore become even more important for Zionist settler ecologies to restore the ancient akkoub and za'atar landscape. Increasingly, the Palestinian foragers are seen as hindering this process and as devastating the contemporary landscape in the region.

Such a "declensionist" perception has been justified through a limited scientific study published in 1995 by an ecologist from INPA who asserted that commercial foraging was likely causing a "significant decrease in the number of flowering *G. tournfortii* in harvest areas."[16] The study warned of "the threat of decline in reproduction and in the number of plants in the long term."[17] But whereas the study recommended restricting only commercial harvesting of this species, INPA imposed a strict ban on all forms of foraging, including domestic foraging. Didi Kaplan, who authored the original research paper, told me that he felt "awful" about how his work was utilized for political ends. "Anyone who knows me knows how far I am from these political ideas," he explained.[18] At the time, he issued a directive in his region that instructed his rangers not to enforce this law against Palestinians who foraged for private consumption. Another high level INPA ecologist asked to emphasize that: "In a situation like this, those who are hurt are individuals and families, while the commercial companies act in a more sophisticated way and obtain an advantage."[19]

Such extractivist capitalist agendas are part and parcel of the settler colonial management of nature. Indeed, over the years, the za'atar was commercialized by both Jewish Israeli and Palestinian farmers, and there are currently numerous farms situated on both sides of the Green Line (the internationally recognized border of 1948-Israel).[20] Because the state has prohibited its foraging for a shorter time and since it is more complicated to cultivate this plant, there are only one or two akkoub farms in the entire region. The first and largest farm is managed by Eli Galilee, an Israeli Jew, in a kibbutz in the northern region of Galilee. During our interview at a roadside gas station near the farm, Galilee told me that the idea of cultivating the akkoub came to him while he was hunting with a few "Arab" friends, who abruptly stopped a wild boar chase when they stumbled upon the akkoub.[21]

> One of the hunters abandoned everything and wandered off toward the other side of the Wadi, like some lunatic. I didn't understand how that could be, in the middle of all the action. My friend [explained that] he found akkoub, and that once there's akkoub, it's better than any meat, or in fact anything else in the world. I said, "Wow, that's interesting," and my research started right then and there. That was 25 years ago.[22]

The process of transforming the akkoub from a plant tied to rich Palestinian culinary traditions to a protected wild herb with domesticated commercial breeds is far from complete. For the most part, Jewish Israelis have not taken to its peculiar taste—a blend of artichoke and asparagus—and it is still

1 I coin and discuss the term "settler ecologies" in Irus Braverman, *Settling Nature: The Conservation Regime in Palestine-Israel* (University of Minnesota Press, 2023).

2 On the power of the imaginary landscape, see Edward Said, "Invention, Memory, and Place," *Critical Inquiry* 26, no. 2 (2000): 184. On the declensionist narrative that blames the native for the desolate state of nature, see, e.g., Diana K. Davis, *Resurrecting the Granary of Rome: Environmental History and French Colonial Expansion in North Africa* (Ohio University Press, 2007).

3 The term "Palestine-Israel" refers to any part of the contested geographic area of the State of Israel's post-1967 territories, including the occupied West Bank, Gaza, and Golan Heights (al-Jawlan). I deliberately chose to use a hyphen rather than the common slash (as in "Palestine/Israel") to highlight the intention to move beyond the bifurcation of this space toward its decolonization. As for the order, it seemed both historically accurate and also more just to place Palestine first. See Braverman, *Settling Nature*, 269.

4 Irus Braverman, *Planted Flags: Trees, Land, and Law in Israel/Palestine* (Cambridge University Press, 2009).

5 Amitav Ghosh, *The Nutmeg's Curse* (University of Chicago Press, 2021), 55, 57–58.

6 See, e.g., Irus Braverman, *Coral Whisperers: Scientists on the Brink* (University of California Press, 2018).

7 David Harvey, "The 'New' Imperialism: Accumulation by Dispossession," *Socialist Register* 40 (2004): 71–90. Patrick Wolfe insisted on the distinction between elimination and exploitation, explaining that the genocidal logic is at the heart of settler colonialism. Patrick Wolfe, "Settler Colonialism and the Elimination of the Native," *Journal of Genocide Research* 8, no. 4 (2006): 387–409. Although many have followed in his footsteps, this distinction is increasingly contested. See, e.g., Sai Englert, "Settlers, Workers, and the Logic of Accumulation by Dispossession," *Antipode* 52, no. 6 (2020): 1647–66.

8 Raja Shehadeh, *Occupier's Law: Israel and the West Bank* (Institute for Palestine Studies, 1985).

9 Alex Levac, "'It Was Nothing Personal,' Bereaved Palestinian Father Told," *Haaretz* (April 4, 2014).

10 Ronit Vered, "How Za'atar Became a Victim of the Israeli-Palestinian Conflict," *Haaretz* (May 7, 2017). See also Ronit Vered, "Forbidden Fruit," *Haaretz* (March 13, 2008).

11 David Lev, "Arab Fined for Picking Near-Extinct Plant," *Arutz Sheva* (June 23, 2013), https://www.israelnationalnews.com/news/169221.

strongly associated with Palestinian culture. On their part, the Palestinians were initially wary of the Jewish commercial breeds. "There are those who said 'without thorns, we won't eat it,' and some told us 'this is not wild akkoub, [so] we won't eat it...It took a while until they understood it's good and just as tasty.'"[23] Today, the commercial breed provides for roughly 20% of the market.

Whereas in most of the ecological juxtapositions I had documented in my research of settler ecologies, the wild organism is associated with the settler state and pitted against a domestic other that is associated with the local enemy, here both the settler and the native associate themselves with the same organism and are therefore battling over its definitions and legitimate uses. On the one hand, settler ecologies want to see a fully wild plant that is not consumed by humans; on the other hand, native practices challenge the wild-domestic divide and treat the akkoub and the za'atar as "nature-culture" hybrids.[24] In this sense, the war that takes place through and over the akkoub is about the proper definition of nature: while the Palestinian communities don't see how their uses of the plant would take away from its wildness, the all-or-nothing approach of Zionist conservation sees it as *either* a plant for human consumption *or* a protected wild plant – but never as both.

The battle over the identity of the akkoub brings us full circle to the problematically dichotomous prism of settler ecologies. It is precisely in challenging this bifurcated context that Adalah's attorney, Rabea Eghbariah, celebrated INPA's recent revision of the law to allow personal foraging of the akkoub of up to five kilograms inside nature reserves and up to 50 kilograms outside of the reserves – in both cases, it should be noted, "without [the akkoub's] roots."[25] The native's interpretation of the plant's identity seems to have won, even if temporarily.

The akkoub story finally highlights a less studied aspect of settler ecologies: the ambivalence of the settler ecologist. Working for the state, the conservation managers and scientists I spoke with were not always fully aligned with its colonial agendas. This was already apparent in the statements I recorded from Didi Kaplan – the scientist whose work served as the foundation for the prohibition on the akkoub. Similarly, Israel's current chief plant ecologist at INPA, Margareta Walczak, has expressed concerns with the dominant narrative of extreme protection promoted by the state. First, she explained in our interview, such an approach is not founded upon sound scientific research. She is currently executing such research but admits that this research, too, is limited in scope and capacity. In her words: "I was all the time very skeptical [about] how much you can really find out regarding the influence [of grazing] because, even in 2005, the plants were already grazed for many years."[26] Walczak is well aware of the political undertones of the foraging prohibitions and worries about the resulting alienation of the local community and how this alienation might in turn undermine broader conservation efforts in the region. She told me along these lines, "The [Palestinians'] connection [to land] is very, very important when you want these people to also care for the land and for the nature around [it]."[27] Walczak blames the antagonistic approach of many local Palestinians toward nature protection on INPA's rigid enforcement practices, which at the same time ignore scientific reasoning. For her, then, the problem is not with conservation science but with law and policy as they manifest in INPA's inflexible enforcement. In her words:

> Who decides at the end? It's more the enforcement unit than us [scientists]. We–and by we, I mean the scientific department–have quite a clear opinion, which is based on facts in the field, that we should allow people [to forage] because it's really part of the culture and tradition.[28]

But while she is sympathetic toward "small" acts of foraging, Walczak at the same time recites the official state narrative regarding akkoub and za'atar conservation, arguing that these plants are in decline and in danger of becoming extinct, that commercial foraging is detrimental to their protection, and that the foraging happens in large quantities and includes massive exports to neighboring countries. For her, then, the challenge is how the state might legally and practically distinguish foraging for personal consumption from that conducted for commercial uses. Relying on fundamentally different assumptions, the Palestinian approach toward the akkoub will typically question each of Walczak's pro-regulatory stances.

The case of the akkoub reveals the ambivalence of the settler ecological actors: although they are part of the state's administration and agree with the fundamental conservation assumptions underlying settler ecologies, these actors are often at the same time acutely aware of the counterproductive and violent repercussions of the myopic focus on nature protection. Such an ambivalent stance by certain state ecologists demonstrates that even at the heart of settler ecologies, possibilities for nonbinary, compassionate, and nuanced perspectives exist that offer more just ecological paths. Ecologies are not inevitably colonial; they can embody and promote coexistence, relationality, and plurality. As demonstrated by the story of "green gold," thinking ecologically might in fact show us the way out of the colonial present toward decolonial ecologies.

12 "Following Adalah Intervention: Israel Reformulating Ban on Harvesting Wild Herbs Used in Traditional Palestinian Cuisine," *Adalah* (August 20, 2019), https://www.adalah.org/en/content/view/9794.

13 Amit Dolev (regional biologist, Northern District, INPA), interview and observations with author, Galilee and Golan Heights (December 23, 2019).

14 Ibid.

15 Jumana Manna, "Where Nature Ends and Settlements Begin," quoting Rabea Eghbariah, "The Struggle for Za'atar and Akkoub: Israeli Nature Protection Laws and the Criminalization of Palestinian Herb-Picking Culture," (Oxford Food Symposium on Food and Cookery, 2020).

16 Didi Kaplan, et al., "Traditional Selective Harvesting Effects on Occurrence and Reproductive Growth of *Gundelia Tournfortii* in Israel Grasslands," *Israel Journal of Plant Sciences* 43, no. 2 (1995): 163–66, 164.

17 Ibid.

18 Didi Kaplan (former regional ecologist in northern region, INPA), Zoom interview by author, February 20, 2021.

19 Dolev, interview.

20 See Jumana Manna, "Where Nature Ends and Settlements Begin," *e-flux journal* 113 (2020); Brian Boyd, "A Political Ecology of Za'atar," *EnviroSociety* (blog) (June 15, 2016), https://www.envirosociety.org/2016/06/a-political-ecology-of-zaatar/.

21 "The insistence on the part of many Jewish Israelis to refer to Palestinians as Arabs rather than Palestinians exemplifies yet another common form of erasure: this time through the refusal to recognize the Palestinian identity." Braverman, *Settling Nature*, 67.

22 Eli Galilee (Jewish Israeli akkoub farmer), in-person interview by author, Kibbutz Ayelet Ha'Shachar, Galilee, June 30, 2022.

23 Ibid.

24 Donna Haraway, *The Companion Species Manifesto* (Prickly Paradigm Press, 2003).

25 Zafrir Rinat, "After Dozens of Years, Akkoub Foraging is Allowed," *Haaretz* (March 10, 2020).

26 Margareta Walczak (plant ecologist, INPA), interview by author, Jerusalem (June 26, 2022).

27 Ibid.

28 Ibid.

Above: A Palestinian farmer, or fallah, serves us traditional za'atar pastries during our visit at his home in Walaje near Jerusalem. Photo by author, February 2018.

Below: Close-up image of akkoub.

XAN SARAH CHACKO

THE VAULT IS A BUNKER, THE ARSENAL ARE SEEDS

Dr. Xan Sarah Chacko is a feminist science studies scholar who writes about the people and practices who are instrumental to science and technology but are rarely credited for their contributions. A lecturer in science, technology, and society at Brown University, Chacko is working on a book about seed banks. She is a coeditor of the recently published *Invisible Labour in Modern Science* (2022) and convenes a monthly speculative fiction book club from her home in Providence, Rhode Island.

+ HISTORY OF SCIENCE, AGRICULTURE

Opposite: Svalbard Seed Vault under construction, November 2007.

Am I shivering because of the cold or with anticipatory nerves?

I stand outside a door, inside a frozen cave, deep within a mountain, in the Arctic. I know it's cold because I can see how my breath wafts up in white clouds around me as it escapes my mouth. I am excited because it has taken a great deal of time, energy, and persuasion for me to be here, and now, finally, it is the moment I have been waiting for. But I am also cold. The five layers of insulating fabrics, two pairs of woolen socks, heat-tech hat, ski gloves, and goggles are doing their best, but my tropical body reminds me that I am out of place. Glistening white crystals cover the walls, ceilings, and pipes that run along the sides of the cave. And save for the recognizable hum of refrigeration, it is eerily quiet.

The black-rock floor also has a sheen and looks like it could be slippery. The deeper I walk into the mountain, the warmer it gets until I reach the door of the freezer. The locked door that lies before me will take me to the coldest place in the mountain. "This is the most valuable room in the world," says a man with a glint of pride in his eyes as he leads the way into the frozen vault. The man, Cary, says that this place came to him in a dream: a dream for the future of humanity.

A tall wool-capped man, Åsmund, has keys to the vault and after a few good turns that awaken the frozen lock, the door is opened. I brace myself for the blast of cold air that manages to penetrate my goggles and sting my eyes, creating pools of tears. Goggles off. My eyes adjust to the cold and soon colors and shapes are sharper and crisper. We walk into a big room. All around us, arranged in neat shelves, are multi-colored labeled boxes: a green plastic box with a sticker that says "ICARDA," a white cardboard box with printed blue letters that read, "United States Plant Germplasm System," a gray plastic box has a punched metal label, "CIAT," a wooden box painted red that has white spray-stenciled letters that spell, "DPR of Korea." These

labels reference institutions and governments from around the world that have sent precious objects in these boxes to be stored for the future in this frozen room, inside an icy mountain, on an island in the Arctic Ocean at 78° North.

I see hope in this valiant effort to save these objects and feel amazed by the providence of the humans who want to believe in a future where these objects will reemerge to save us all. At the same time, I see the despair in the idea that these objects may survive even if the world outside is ruined. I see desperation for at least something to persist against the odds of survival. I also see variation in the colors of boxes, the materials used to make and mark the boxes, and the difference in the number of boxes from some institutions and countries versus the others. I see omissions, gaps in my mental maps of countries, big and small, that are missing from representation in this future. At some point I stop seeing things in the frigid warehouse because my body has had as much as it can take of the cold. My face is frozen. I need to leave the future.

Bunker Logic

Seed banking–done by taking the reproductive parts of plants and placing them in vaults (usually frozen)–is a technoscientific endeavor that has gained prominence in the late 20th century as the foremost means of securing the future of plants. One of the evidentiary examples of the human-made planetary crisis in climate is the mass extinction of plants, which is often portrayed as an event but is better understood as a structure. Seed banks are boundary objects–demarcating the boundary between life and death–that at once serve as a reminder of human exploitation of nonhuman life and at the same time have come to embody collectivized hope for human survival. The history, theory, and practices of seed banks are articulated through an omnipresent militarized logic. Seed banks are created in the image of bunkers, while the seeds they hold are imagined as arsenal against the threats of climate change, food insecurity, and the biggest threat of all, ourselves. What makes cryopreservation different from habitat restoration or rehabilitation is that the seed takes the form of a promise of life in the future. Seeds are particular in their capacity to regenerate after long periods of stasis, after all, they do this on their own. They are living beings, and yet in seed banks their vitality is suspended, thus making them more like specimens in museums. But how can we know our investments will come to fruition? What guarantee do we have that these seeds will thrive in the future world they inherit?

Working on a project about the history and practices of seed banks, I noticed in the mid-2010s that something had changed about the public awareness of this technoscientific enterprise. A decade earlier, when I started work on seed banks, friends' and family's eyes glazed over when I started talking about my research on the care practices of seed curators. The idea that to keep plants alive we should freeze their seeds seemed puzzling to folks for whom conservation meant protecting a species' capacity for life. This meant either saving environments or helping certain species overcome hurdles to reproduction to increase their numbers (think of how individual panda births in zoos quickly become global phenomena). Ecologically minded friends and colleagues understood that saving plants was as important an endeavor as the visible movements to conserve mammals and sea creatures. However, the concept of the frozen vault of seeds as a bank for the future required the additional layer of the science of cryopreservation, which isn't too far from the now commonplace act of freezing eggs and sperm. Thinking of seed banks within the context of

human tissue biobanks brought the project out of the realm of maintaining life to creating new forms of it.

Something had changed in the cultural milieu and those same friends and family were suddenly wide-eyed with interest and proceeded to send me articles and podcasts about the future of the world's plants and the imbrication of seed banks in that future. The iconic edifice of the Svalbard Global Seed Vault (SGSV), known by its click-baity moniker of "doomsday vault," had invaded the collective imaginary of well-intentioned readers of The New York Times and The Guardian. While I read the coverage of the SGSV in the English language news to understand the way that it was perceived, I noticed that most articles fell into two categories – ones that propose the vault as the hope for humanity, and ones that use it to despair in our ruination of the earth. All coverage agreed on the urgency to conserve plants in the face of widespread loss of agrobiodiversity and the consequent panic about what this means for the future of humanity. Newspaper articles, magazines, journals, and websites all respond in cycles of hope and despair using Svalbard as a symbol of the apocalypse. As an object that so evidently straddled the boundary between hope and despair, SGSV opened for me a vector through which the entire project of seed banking could be understood.

The Northernmost Freezer in the World

The SGSV is located outside the city of Longyearbyen, in the Svalbard archipelago. A treaty recognizes Norway's claim to the land but allows for science and exploratory mining to be conducted by other signatories such as Russia, the United Kingdom, and the other Scandinavian countries. Named for the United States coal mining magnate, John Munro Longyear, the town has first been a site of extractive coal mining since the early 20th century. Longyear, representing the interests of the Boston, Massachusetts-based Arctic Coal Company, visited Svalbard in 1901, bought the claims to mine in 1903, and subsequently established the town of Longyear City as a mining hub. When you visit Svalbard, the traces of the mining industry linger in the most prominent landmarks – the coal miner's cabin is a popular place to stay, and the aerial tramway to move the coal forms a backdrop along the mountains that surround the fjord.

Citing financial difficulties during the First World War, the Arctic Coal Company sold its mining interests to the mining company Store Norske. While Longyearbyen continues to be the most populated town on the archipelago, tourists are encouraged to visit the neighboring Russian towns of Pyramiden and Barentsburg. Like its southern correlate in the Antarctic, life in the Arctic Circle is surprisingly cosmopolitan – Longyearbyen has not one, but three Thai restaurants, and its whiskey bar, The Karlsberger Pub, boasts nearly 400 different kinds of whiskey. And a good thing too; where else is the researcher visiting at a time of year when the sun sets at 2pm supposed to go when it's dark?

It was an abandoned Store Norske coal mine that provided the space for the first deposition of seeds in Longyearbyen, as part of a longitudinal study called "the 100-year project." This longevity test of the permafrost's capacity to store seeds for the long term was initiated in 1986 by the Nordic Genetic Resource Centre (NordGen). Scientists selected 42 seed samples of the 16 most important Nordic plant species and placed them in canisters deep in the abandoned coal mine. Seemingly unrelated, a year later, the Nuclear Waste Policy Act of the United States of America was amended to designate

Yucca Mountain, Nevada—a sacred place for the Shoshone and Paiute people—a nuclear waste repository. While in Svalbard the permafrost was being hailed as nature's workaround for the need to artificially cool seeds to slow their decay, in Nevada, the erasure of the Indigenous people from the minds of policy makers rendered the mountain empty and safe – safe enough to allow the nuclear waste to decay. While the toxicity of the nuclear waste buried in one mountain demands 10,000 years of no contact, how odd that in another mountain even in our most experimental and hopeful imaginings of the future, 100 years is all they could muster for keeping plants alive.

In the first 20 years of the 100 years, parts of the seed samples were assessed for their germination and viability every two-and-a-half years and thereafter they are being checked every five years. The results of the first 20 years of testing on the seeds stored in permafrost showed that the permafrost is warming, and the seeds need to be cooled much further than -4°C or -6°C. When the plans for the Global Seed Vault were being made, it was already known that the rooms that would house the seeds would need to be cooled further with refrigeration supplied by electricity, made from coal, from this mountain, grossly subsidized by Norway. The mountain was excavated and the biggest freezer closest to the Arctic was built and ready for deposition of seeds by 2008.

The Vault in the Media

The first article that references Svalbard is from a 2006 *New York Times* reprint of a Reuters bulletin titled, "NORWAY: SAVING SEEDS, AT 40 BELOW ZERO."[1] The article announces the plans for a "Global Seed Vault to protect frozen seeds—for crops like rice, wheat, barley and fruits and vegetables—from cataclysms." In its short summary, the article mentions the plans by the Norwegian government to establish a vast ark for seeds, explicitly invoking the metaphor of Noah's Ark, and describes the location as "600 miles from the North Pole." The initial coverage of Svalbard in the news media mobilized rhetoric of hope and salvation. Two headlines evidence this trend: "AFTER THE APOCALYPSE: INSIDE THE ARCTIC VAULT THAT COULD HELP KEEP HUMANITY ALIVE," and "THE DOOMSDAY VAULT: THE SEEDS THAT COULD SAVE A POST-APOCALYPTIC WORLD."[2] It is notable while that the word "apocalypse" shows up in both cases, the SGSV is not yet portrayed as the definitive solution. The use of the word "could" rather than "will" indicates possibility but not certainty. The implication is that only after the apocalypse will it be revealed whether this investment saves us or if it is a waste of resources. Other news stories from the early days of the vault are more skeptical with headlines like, "Why seed banks aren't the only answer to food security," and others calling it "simultaneously one of the most forward-thinking endeavors humans have ever undertaken – and one of the most obscure."[3]

And yet others probe the funding of the vault and claim its political possibilities are downright diabolical. The article "Bill Gates and GMO Cronies Plan $30 Million Seed Vault While Poisoning the Planet" reminds readers of a particular connection that, on the one hand, the SGSV is funded partially, via the Crop Trust, by the Gates Foundation, but that, on the other hand, the Gates Foundation also held shares of agribusiness giant Monsanto, which is now owned by multinational pharmaceutical and biotechnology giant, Bayer.[4] The insinuation is that by owning their shares, the Gates Foundation was complicit in the nefarious activities of Monsanto, and thus any investment in the SGSV taints it with the same brush of complicity. The author Christina Sarich writes, "If you think the

recent scandals of the Wall Street Banks, Libel, and the Federal Reserve were outlandish, imagine global bio-warfare on the world population. That is what this is leading to. It will make the Russian bread lines flanked by the KGB of times past look like a play date."[5] Sarich goes on to state that the only possible way for us readers to resist this iniquitous global plot is to start saving seeds and planting vegetable gardens in our lawns. The ferocity with which SGSV mobilizes our anxieties about the future is clear – it could just as easily save or destroy us.

Seeds as Weapons

Sarich may have been inspired by the 2009 science fiction novel, *Windup Girl*, by Paolo Bacigalupi, where control over seed banks changes the fate of the world. Set in post-apocalyptic Thailand, where the national seed bank is hidden and possibly mobile, the Trade Minister is speaking to Anderson Lake, a representative of AgriGen, an agribusiness giant who has control over the major global crops after having secured their intellectual property (not unlike Monsanto). The minister says,

> The seedbank has kept us independent of your kind. When blister rust and genehack weevil swept the globe, it was only the seed bank that allowed us to stave off the worst of the plagues, and even so, our people died in droves. When India and Burma and Vietnam all fell to you, we stood strong. And now you come asking for our finest weapon...No farang [foreigner] should ever touch the heart of us. You may take an arm or a leg from our country, but not the head, and certainly not the heart.[6]

The seed bank is the heart of the country because it protects the most valuable treasure, the cultural and natural heritage.

Previous Left: The Tunnel in Svalbard Global Seed Vault. Longyearbyen, Svalbard.

Previous Right: The underground Exploratory Studies Facility, Yucca Mountain, Nevada.

Opposite: Inner Door, Svalbard Global Seed Vault, Longyearbyen, Svalbard.

Above: Ground level rendering, The Arc – a Visitor Center for Arctic Preservation Storage in Svalbard.

Next: Seeds from the Navdanya Farm, Derhadun, India.

But why are plants under such threat and needing to be saved in the first place? A mechanistic understanding of evolution pits species against each other for survival. In this neo-Darwinian Hunger Game plants are actually doing quite well. Plants still outweigh animal biomass 400 times over. For all our anthropocentrism, plants have been gaming natural selection in more ways than we can imagine. They have evolved relations where they reward animals for protecting them from herbivore predation, they communicate to family members about oncoming threats of predation, and they are able to synthesize toxins specific to their attackers. Theorizing life as war makes plants generals of strategy, cunning, and violence. So why do they need protecting? Why should we be worried? Could it be that the fallout of our militarized approach to understanding plant life and responses in agriculture kills rather than protects?

Feminist and farmers' rights activist Vandana Shiva provides a prominent critique of industrialized agriculture that the crops put forward through the system do not always provide the salvation they promise and instead bring with them ecological ruin and financial dependence.[7] Virulent pests led to the use of stronger pesticides and demands for higher yields meant more fertilizer use. The combination of heavy use of chemical fertilizers and pesticides resulted in two problems for the future of small-scale farming. First, chemical runoff of pesticides and fertilizers was poisoning soil and water systems and leading to environmental degradation; and second, the dependence on the agricultural industry for seed, fertilizer, and pesticides created a system of debt and required a capital investment that many subsistence farmers did not have. The desperation and struggle of these communities is evidenced by the wave of over 127,000 farmer suicides in India in a little over a decade that were linked to just one crop – genetically modified cotton.

In 1987, the same year as the Yucca Mountain decision and just one year after the 100-year project had started in Svalbard, Shiva founded Navdanya Biodiversity Farm, a farmers' collective to save and distribute indigenous seeds across 22 states in India. According to their origin story, Navdanya, which can be translated to "nine seeds" or "new gift," was a response to both the farmer suicides and the Bhopal gas leak of 1984, where industrial producer of pesticide Union Carbide exposed 500,000 people to methyl isocyanate, an intermediate chemical in the production of carbamate pesticides. Shiva juxtaposes the tragedies of willful suicide of the farmers, and the mass death and mutilation associated with the unseen and immediate killer of methyl isocyanate. In both cases data had to be aggregated over years to see the scale of the devastation to human life.

Contemporaneous in their start, Navdanya and the 100-year experiment envisioned completely different projects that were responding to the same anxieties about crop diversity loss, degradation of farmland, and uncertainties about the future. However, only one routinely makes global headlines, occupies the public imaginary of conservation, and continues to deploy militarized logics of hypersecuritization through bunkers. Elsewhere I write about how the future is made in seed banks through layers of alienation and reinscription and how the concepts of "biodiversity" and "natural resources" continue the colonial practices of commodification and extraction.[8]

For now, I conclude with the chilling report from *Fortune* magazine that Martha Stewart is not worried about the apocalypse hitting tomorrow because she is confident that we have prepared.

> Instead of stockpiling canned goods or dried pasta, she's putting her faith in the Svalbard Global Seed Vault. Think of it this way: The Global Seed Vault is like the cloud, storing up to 4.5 million backups of heirloom (non-hybrid) seeds ranging from African wheat and rice to European eggplants and potatoes. If global warming destroys all the corn in the Midwest or the avocados in Mexico, we'll be able to access those seeds and restart crop populations without them going extinct.[9]

Ms. Stewart made a reconnaissance visit in February 2018 when she led a tour to the SGSV to meet its founder Cary Fowler, while enjoying Michelin-rated dinners, a champagne reception to view the Northern Lights, and even a day of adventures including a polar bear tour, dog sledding, and exploring a glacial cave. For a small donation of $10 to the Crop Trust any of you could have been the lucky two to accompany her.

1 Reuters, "Norway: Saving Seeds, at 40 Below Zero," *New York Times* (May 31, 2006), https://www.nytimes.com/2006/05/31/world/europe/31briefs-brief-002b.ready.html .

2 Arwa Damon & Gul Tuysuz, "After the Apocalypse: Inside the Arctic Vault That Could Help Keep Humanity Alive," *CNN* (October 26, 2015); Suzanne Goldenberg, "The Doomsday Vault: The Seeds That Could Save a Post-Apocalyptic World," *The Guardian* (May 20, 2015).

3 Simran Sethi, "Why Seed Banks Aren't the Only Answer to Food Security," *The Guardian* (November 26, 2015); Chris Mooney, "Why the World is Storing so Many Seeds in a 'Doomsday' Vault," *Washington Post* (April 15, 2016).

4 Christina Sarich, "Bill Gates and GMO Cronies Plan $30 Million Seed Vault While Poisoning the Planet," *Waking Times* (July 3, 2013).

5 Ibid.

6 Paolo Bacigalupi, *The Windup Girl* (Night Shade Books, 2012), 151.

7 Vandana Shiva, *Monocultures of the Mind: Perspectives on Biodiversity and Biotechnology* (Palgrave Macmillan, 1993).

8 Xan Sarah Chacko, "Creative Practices of Care: The Subjectivity, Agency, and Affective Labor of Preparing Seeds for Long-term Banking," *Culture Agriculture Food Environment* 41 (2019): 97–106; "Stringing, Reconnecting, and Breaking the Colonial 'Daisy Chain': From Botanic Garden to Seed Bank," *Catalyst: Feminism, Theory, Technoscience* 8, no. 1 (2022): 1–30.

9 Anne VanderMey, "This is Your Last Chance to Visit the 'Doomsday Vault' and Go Dogsledding with Martha Stewart," *Fortune* (January 25, 2018).

ANDREA LING
DESIGN(ED)

DECAY

Andrea Ling is a researcher working at the intersection of design, fabrication, and biology. Her work focuses on how the critical application of biological and computational processes can move society away from exploitative systems of production to regenerative ones. She is the 2020 S+T+ARTS prize winner for her work as the 2019 Creative Resident at Ginkgo Bioworks designing the decay of artifacts in order to access material circularity. She graduated from the MIT Media Lab, Mediated Matter group, where she was a research assistant on the Aguahoja project. Andrea is an architect with the Ontario Association of Architects and a founding partner at designGUILD, a Toronto-based art & design collective. She is currently an A&T fellow at the Institute of Technology and Architecture at ETH Zurich working on living material synthesis.

+ ART, ARCHITECTURE, BIO-DESIGN

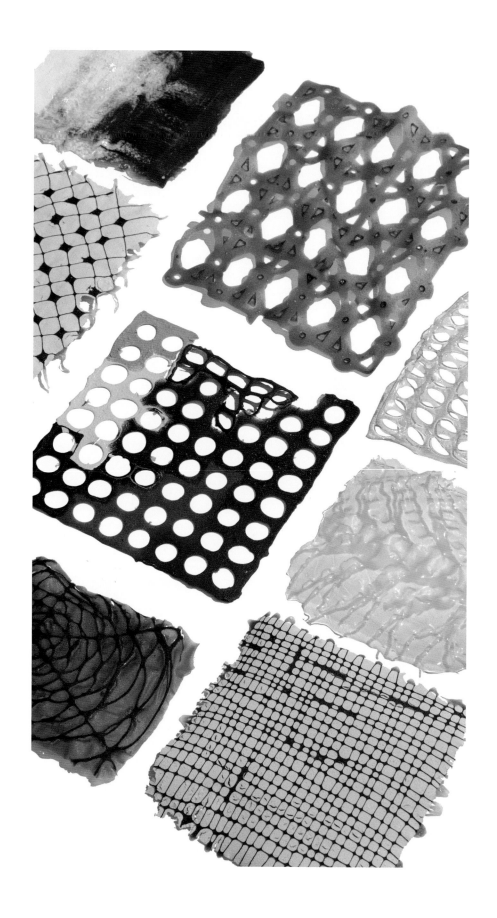

The Gwion Gwion rock paintings of Western Australia are estimated to be 46,000 to 70,000 years old. Found in areas exposed to rain and sunshine, the paintings remain vivid despite never having been retouched. The story they depict is not my focus – instead, of interest is the paint that was used. Derived from plant material, the original paint was decomposed by a symbiotic community of black fungus and red cyanobacteria that, due to their growth conservative nature, stayed within the confines of the original drawing, each successive generation of microbes cannibalizing previous generations *in situ*, and rendering the paintings in a living pigment that has endured far longer than any human-made material system. These paintings endure because they materially die and renew, finding longevity through transformation rather than static endurance.

These paintings are reflective of how nature makes decomposition useful and serve as inspiration for a series of projects where I use the intentional decay of biologically derived material, much of it from botanical sources, as a constructive force. By organizing decomposition through controllable parameters–such as type of decay agent, spatial templating, and control of the decay environment–these projects seek to simultaneously deconstruct and reconstruct matter such that the formed objects are completed by changing. In this manner, decay is a fabrication process, using biological agents on biologically derived materials to transform the artifacts as well as exploring how mechanisms of constructive renewal can be built-in to the objects we make.

Design by Decay, Decay by Design is a series of artifacts expressly designed to decompose. Done for the synthetic biology company Ginkgo Bioworks as part of their 2019 creative residency, the artifacts are impregnated with enzymes, bacteria, and fungus that erode material away in specific fashion. The objects are made of biologically derived and biodegradable composites of chitosan, cellulose, and pectin, some of the most abundant biopolymers on the planet, harvested from waste from shrimp farms, wood pulp mills, and agriculture waste. All are water soluble, with short decay cycles, and can be extruded and cast as water-soluble colloids. The gradation and proportions of these three ingredients can be tuned to create a vast design space whose characteristics include opacity, color, flexibility, and mechanical strength. Only three structural polysaccharides–minimally processed–ensure these characteristics are achieved, all the while easily accommodating biodegradation. That is, a whole artifact is made of a small library of ingredients that do not require vastly different end-of-life processes to decompose and can be easily taken up again into a new nutrient cycle. This is in direct contrast with many industrial material systems whose complex assemblies require great effort to separate their constituent parts before being taken to different recycling facilities, scrap yards, and landfills for disposal.

Design by Decay, Decay by Design, is comprised of three smaller projects, each of which asks how one can harness the responsivity and temporality of biologically derived materials in order to shape new things:

The first series used enzymatic degradation [opposite page]. Cocktails of cellulase, pectinase, chitinase, amylase, and a general lysing enzyme were applied in different concentrations and on different compositions to test their effects, with the goal to use enzymatic degradation as an *in situ* fabrication tool. The enzymes were used to selectively degrade holes and lines in pectin, chitosan, and cellulose skins during the casting process, subtracting material from the artifacts as they were formed. As a control, a large pectin-cellulose-chitosan "painting" was laser cut and dried, and its resolution compared with a similar painting where enzyme was used in lieu of laser cut lines. The enzyme degraded lines were of poor resolution–a function of the enzyme's diffusion through the semi-wet material–and often ate away more material than expected, leaving powdery residue in its wake. Interestingly, degradation continued even when the colloids were dry to the touch, seen as the lines and holes became larger with time. The most compelling results were obtained when using the enzymes to create holes and perimeter lines for a series of tripod structures. Here, the material was successfully transformed through degradation into similarly sized objects that, though not as standardized as industrially produced objects, were similar enough that they could be aggregated with some modularity.

The second project involved carving and coloring cellulose forms with different types of fungi. *Aspergillus niger* is a common black mold found in soil and on fruits and vegetables. It produces pectinase and amylase. *Trichoderma viride* is a common green mold found in soil that produces cellulases and chitinases and can be used as a bio-fungicide against other plant pathogenic fungi. Both are common and effective decomposers of biodegradable material. *Aspergillus* was templated onto cellulose-chitosan composites with stencils, growing within and sometimes overspilling the confines of the template and eventually eating away material according to the pattern. Co-cultures of *Trichoderma* and *Aspergillus* were also grown onto carved blocks of maple, milled to quadruple the surface area available for colonization. Inoculated onto different areas of the wood, both organisms quickly intermingled spatially on the blocks, and sometimes became "contaminated" with a third species of white mold. Along with coloring the blocks with their vibrant green, yellow, and black pigments, touch tests showed the blocks softening with time as the cellulose was broken down.

The third project involved coloring cellulose and chitosan-based objects with *Streptomyces*, a genus of filamentous bacteria that is found predominantly in soil and decaying vegetation. They are researched as a major source of the world's antibiotics and produce geosmin, a metabolite that

gives soil its characteristic earthy rain smell. Many species in the genus also produce vivid pigments as metabolites, which can be used as alternatives to industrial dyes for textiles since they use much less water. Different strains of *Streptomyces* were cultured on small samples of wood, chitosan-cellulose composites, and rice paper to test if they could dye these samples while gently decomposing them. The intent was to use the bacteria to color a chitosan-cellulose cocoon structure, while also eroding it. Scaling up to this larger structure, however, proved exceptionally difficult. While we knew *Streptomyces* can produce cellulases and chitinases, they are bacteria that can be easily contaminated and overwhelmed by molds and sterilizing the biomaterial composites without damaging them or producing toxic residue was an issue. Initial strip tests, incubated over 30 days, resulted in the growth of a white mold; however, there was no evidence of the *Streptomyces*. Attempts to grow the bacteria on thick, sterile maple blocks also resulted in heavy contamination. In one instance, the results smelled so vile it had to be discarded; in another, a white mycelium-like texture resulted; and, in yet another, evidence was found of a fuzzy brown-gray contaminant on top. Herein lies some of the irony of doing this work in the controlled setting of a biology lab rather than simply burying samples in exterior soil. These were wild strain bacteria directed to do what they naturally would do outside, on a dead tree stump, decaying leaves, or rotting fruit, with other micro-organisms present and thriving. However, inside a lab's highly artificial conditions, which demand sterilization for growing predominantly monocultures that are then sensitive to contamination, this control seemed to hinder rather than enable the growth of this bacteria and the controlled decay of a structure.

The work from Ginkgo Bioworks highlighted some of the issues that arise when artists and designers try to work with biological matter at scales and environments beyond the confines of a wet lab. In all the tests, "contamination" was always a factor. As someone who wants to design the decay of these artifacts in a specific way, I must reflect on whether these accidents are just as effective as planned forms of decay, and what my priorities are for these artifacts as they change – are they aesthetic, sensorial, or programmatic? These factors all must be weighed, highlighting how bio-designers, as well as scientists, are called upon to decide when it is necessary to guide biology with a firm hand and when it is better to let go. As a classically trained architect, I am used to prescribing not only the aesthetic quality of a work, but also the performance and sequence of how things are assembled, sometimes to submillimeter tolerances. It is very difficult to relinquish control and outcomes to these natural partners. But, in the end, all fabrication tools, including industrial ones, produce results that are always an approximation of the original design, and that it is out of the interplay between intent and outcome that some of the most interesting processes and results reveal themselves.

These ideas were expanded on in *Calculus of an Infinite Rot, Part 1*. Commissioned by Toronto's Rhubarb Theatre Festival in 2022, the project is an attempt to take templated and larger scale decay processes outside of the wet lab. Thirty-four tree stumps collected from fallen maple and spruce trees were carved to varying levels of finish, inoculated with different types of benign fungus and bacteria, and left to incubate for four to six weeks in the theater. The rotting stumps were then installed on stage as a living container for various performance artists to respond to regarding how they might regenerate their artistic practices after two years of pandemic isolation. What would the performers leave to deteriorate and what is used as fodder and fuel for new creation? This was asked in the most literal sense, as the stage set was actually rotting, employing millions of micro-organisms as an assist. The artists were asked to reflect on the impact of time and entropy on their production, and to consider that decay and regeneration are paired processes where the disintegration of one system is used for the organization of another – in a circular ecology, one cannot have one process without the other. The title, *Calculus of an Infinite Rot*, refers to Iranian philosopher Reza Negarestani's position that decay is not the marked absence of life or wholeness, but is rather the negotiation between shifting states of living and dying. Biological systems are inherently dynamic, and processes like growth and degradation rely on a temporal flow of energy and material exchange between organisms. Through decay, the differentiation between humans, trees, and microbes blurs, as parts of us become parts of them, and vice versa, in a process that is leaky, smelly, and messy.

It was intentional that the objects that were sculpted to rot for this piece were made of wood, a material that, in the living tree, is already dead. That is, in living trees, the vast majority of the tree's cells—outer bark, heartwood, and sapwood—are dead. These cells act as protection, structural support as the tree grows, or transportation conduits for water and nutrients to flow between leaves and roots. The tree requires its dead cells to support its living ones, and without cell death the tree cannot grow. Here is nature's insistence that material is not wasted, not only with the potential to be "upcycled" as compost when the wood finally does decompose, but also to be employed in the dead state to facilitate growth of the living.

In my first year of architecture school, as part of our (Western) cultural history classes, we studied the *Epic of Gilgamesh* as one of the first examples of city building. King Gilgamesh of Uruk, with the assistance of the savage Enkidu, slays Humbaba, a monster that protects the sacred forest. With Humbaba dead, Gilgamesh cuts down the sacred cedars and uses the lumber to build gates for his city walls. Gilgamesh and Enkidu then slay the Bull of Heaven; however, for this, Enkidu is killed as punishment by the gods. When Enkidu's body starts to rot, Gilgamesh, fearful of this natural reminder of mortality, casts the corpse outside of the city walls, making a statue of Enkidu inside the city to immortalize his friend.

Gilgamesh is the first great builder who, terrified of the uncontrollable wilderness, builds a city with tall walls to tame his surroundings, and separate culture from nature–the built environment away from the botanical one–in the belief that he can build his way to immortality if the statues are tall enough, the walls are big enough, and the dead are outside. Six thousand years later, we see what happens when we approach the world in this way. When we construct our world based on Gilgamesh's extractive legacy, one that rejects the role of the protectors-monsters of the wild and the dead within regenerative cycles, we invariably end up accidentally destroying the mechanisms of resilience and renewal that are often paired with these monsters. Without these mechanisms, the city is not so much human's vessel of immortality that Gilgamesh wished it to be, as it is a tomb. Integrating rot into these "human-made" projects is not only to confront the qualities of decay that make us uncomfortable but also to invite some of the monsters back inside, along with their potentially restorative processes.

The temporal and physical scales at which decay occurs may not always be perceptible by us–both microbial time and geological time are challenging to understand–but it has tangible consequences and it is necessary to facilitate regenerative modes of production. The use of biologically derived materials is ancient; however, modern use of these materials is challenging. Unlike mineral and petroleum-based materials, biological materials tend to be environmentally responsive over time, fluctuating in their dimensions, water content, color, and other physical properties. They are difficult to standardize and cannot be controlled in the same way we control industrial materials. They do not offer equivalent performance characteristics as conventional material systems and cannot be used as replacements. What they do offer, however, is biocompatibility, resilience, and dynamic capabilities not possible with inert and environmentally agnostic material systems. The variability that makes them so difficult to work with in an industrial context is the same quality that makes them able to respond, repair, and replicate. To partner with these materials effectively requires a different process of design, one that accommodates variation, acknowledges material agency, and pursues mutability as a desired quality in the built world.

It is in this context that I situate my interest in designing decay as a means of accessing new modes of production. By designing with and for decay, we can return material and energy back to the biological system and perhaps bias its outcomes so that we may better live with its consequences. By mediating the decay process through species selection, control of environmental conditions, and templating of nutrients, we can actively pursue self-renewable forms and stewardship of the physical world, as well as guarantee that mechanisms of constructive renewal are embedded in this world.

Acknowledgments

For *Design by Decay, Decay by Design* I would like to thank the curatorial team at Ginkgo Bioworks + Faber Futures–Natsai Audrey Chieza, Dr. Christina Agapakis, Grace Chuang, Kit McDonnell, Dr. Joshua Dunn–and the scientific advisors at Ginkgo Bioworks – Dr. Joshua Dunn, Dr. Ming-Yueh Wu, Kyle Kenyon, Duy Nguyen, Dr. Lucy Foulston.

For *Calculus of an Infinite Rot, Part 1* I would like to thank Rhubarb Festival Director & Curator, Clayton Lee; fabricators Leah Ataide and Nicholas Hoban; and technical support Betty Poon, Anna Gregorczyk, Olga Chomiak, and Natasha Christie-Holmes at Temerty Faculty of Medicine at the University of Toronto. Funding was provided by Canada Council for the Arts & Rhubarb Festival.

p. 52-53: Fragment from *Calculus of an Infinite Rot, Part 1*.

p. 54: Pectin-chitosan-cellulose composites, *Design by Decay, Decay by Design*.

Previous: Fungal wood block, pectin-chitosan-cellulose samples, *Design by Decay, Decay by Design*.

Opposite: Chitosan-cellulose composite cocoon structure, *Design by Decay, Decay by Design*.

IN CONVERSATION WITH

GIOVANNI ALOI

Giovanni Aloi is a curator, author, and educator whose work focuses on representations of the natural world, particularly as they relate to the use of plants and animals in contemporary art. He is founder and editor in chief of *Antennae: The Journal of Nature in Visual Culture* and author of numerous books, including *Speculative Taxidermy: Natural History, Animal Surfaces, and Art in the Anthropocene* (2018), *Why Look at Plants? The Vegetal Emergence in Contemporary Art* (2019), *Lucian Freud – Herbarium* (2019), and *Posthumanism in Art and Science* (2020). Aloi spoke with **Karen M'Closkey** about the importance of art for providing and provoking an expanded engagement with plants.

+ In 2006 you founded *Antennae: The Journal of Nature in Visual Culture*, which just published its 62nd issue. Early on, a majority of the issues were focused on animals. Then, in 2011, you published two issues titled "Why Look at Plants?" and, in 2020, three issues were dedicated to "Vegetal Entanglements." Have you seen significant changes in either the number of artists engaging with plants or in the reception of this work by institutions or publics in the decade between "Why Look at Plants?" and "Vegetal Entanglements"?

Yes, plants are no longer taboo in the humanities. Hurray! It has taken far too long for matters to shift in this direction. I think it's a bit embarrassing, really. I care to say, however, that, in this case, as it was for animals, artists were ahead of scholars in breaking the mold. Artists like Helen and Newton Harrison, Joseph Beuys, Mel Chin, Agnes Denes, Alan Sonfist, and Lois Weinberger, just to name a few, had engaged with plants in new and critical ways. However, institutions and publishers insisted on considering their work within the context of land art rather than embracing the idea that plants played active-agent roles deserving of our attention. Art historians and critics are often limited in what they can say by institutions – the rejected book proposal that was 10 years ahead of its time, or the essay ripped apart by peer reviewers who were not familiar with the ontological turn. Institutions are relentlessly behind—not only are they caught up in bureaucratic lethargies of all kinds, but I have come to believe that they, by essence, tend to be conservative even when they claim to foster progressive and inclusive views—too afraid to risk the status quo or held back by trustees who know little about contemporary art.

+ How has your own focus changed in the decade between these two sets of publications?

I have considerably moved away from the limited optics that characterized what I have come to see as the "late-stage of animal-studies" and I have become much more involved in posthuman, OOO, new materialist, and anthropogenic discourses entailing holistic and diverse human-nonhuman networks.

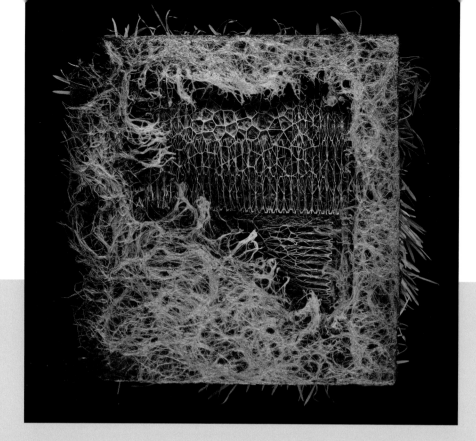

Rootbound #3, Exercises in Rootsystem Domestication (2018) by Diana Scherer.

+ Can you also point to a significant difference in terms of how artists are working with plants today than, say, 20 years ago?

The past 20 years have seen a major shift in art education. The most influential ideas have come from philosophers, anthropologists, and ex-scientists-cum-theorists. As a result, the perspectives that artists are interested in are no longer those of postmodern art theory, and thank goodness for that. I don't think that the self-absorbed art theory of the 1970s has any place in art-making today. Artists are looking at the world with a sense of urgency. They are very critical of institutions as well as past theoretical ideas that excluded women, BIPOC, and LGBTQA+ thinkers and makers. Climate change and social justice have acquired considerable traction and this urgency directly informs the way artists think and work. But perhaps even more importantly, the rise of #metoo, Black Lives Matter, and the perspectives of Indigenous knowledge have led to an extremely interesting and challenging cultural expansion. Western philosophy is now only one of the possible theoretical frameworks at the disposal of artists – the field is more inclusive, diverse, and experimental.

+ You began work in animal studies but quickly grew critical of its approach – you thought it replaced anthropocentrism with zoocentrism due to its focus on sentience as a measure of value, which diminished plants and other organisms that we cannot as easily empathize with. Has the emergence of critical plant studies and various new materialisms generally steered clear of this hierarchical ordering of human, then animal, then plant, or are we simply working our way down the "food chain"?

I don't think it's a matter of working our way down the food chain as much as it is a matter of progressively coming to terms with deeper conceptions of otherness/alterity that were previously impossible to seriously take into consideration mostly, again, because institutions were not ready for it. Every time we focus on a new and more cryptic *other than human group*, we challenge ourselves to redefine our perceptual and empathic registers. I am glad this is happening.

+ The diversity of authors and projects in your book *Why Look at Plants?* answers the question that your title poses, which is that we need many, and varied, ways to engage plants beyond their objectification. On the one hand, the philosophical project of both plant and animal studies is to level the hierarchy of human, animal, plant, and so on, but art is inescapably directed toward human sensibilities. Can you speak about how this philosophical flattening has impacted the representation of plants in art?

I think that it would be incorrect to suggest that the ontological flattening we have witnessed during the past 20 years has come exclusively from philosophy. As I have argued in a paper that I gave in 2015 called "Registering Interconnectedness," artists anticipated flat ontology as early as the 1920s. Joan Miró's constellations, Elsa von Freytag-Loringhoven and Marcel Duchamp's readymades, Hannah Höch and Claude Cahun as well as Dada and Surrealist assemblages. I do not want to generalize but the true baseline of contemporary ontological flattening is agency – the acknowledgment that everything has agency and that there never is only one active (human) subject.

+ This question of agency reminds me of an interview where you said that plants are often presented as benign or cast as a model of community, a stereotype that you referred to as "toxic positivity." You said that those who advocate these benign views do not engage with plants themselves. Are there works that you think provocatively engage this aspect of plants?

Anyone who gardens or works in the agricultural field knows that plants can act in very self-interested ways. Yes, plants can also support each other and establish symbiotic relationships among themselves, animals, and other organisms but anthropomorphic interpretation of these complex relationships can skew our perspective. Toxic positivity is a form of cultural fragility: a compensatory response instilled by the anxiety and loss of hope that results from constant exposure to negative news linked to climate change and the sixth extinction. Over the past 10 years we have witnessed a rise in interest for storytelling as an alternative epistemological and dissemination tool. While there certainly is a productive place for storytelling in multidisciplinary research of all kinds, in the hands of capitalism, far too often, storytelling demands a happy ending. The insistence of casting nature as a benevolent model is extremely problematic because it implicitly attempts to validate the goodness of humans as natural beings. This has happened already with animal studies, and it is now happening with critical plant studies – it is epistemologically unethical. The "nature best sellers" on today's mainstream market are often feel-good books that focus on what we humans perceive as virtue. This is a regressive and totalizing form of inherent anthropomorphism that ultimately damages the progress made during the past 20 years in the humanities. Plants can be awful to each other. They smother and choke, parasitize and poison each other. What we do with that knowledge is the challenge we have to embrace and address. Ignoring these traits simply because they appear undesirable to us, or because we fear that they might promote lack of empathy among humans, is unethical and dangerous since it contradicts true scientific knowledge.

+ Botanic art—specifically images of plant displays and specimens—has long been a genre in art. In the Western canon this would include the rise of still life painting in the 17th century but also modern and contemporary artists, such as Warhol's flower screen prints, O'Keefe's paintings, Mapplethorpe's photographs, and more recently digital drawings by Macoto Murayama. These are just a few examples from a profusion of artists whose depiction of plants focus on flowers as specimens without context. Is there someone working in this genre today who you find compelling in their ability to work within that tradition but also shift our perception in terms of what is being objectified, or symbolized, or perhaps even in the technique of producing the work?

Zachari Logan is a wonderful example of an artist who productively straddles the line between tradition and critical contemporary perspectives. His work harnesses the lush beauty of Dutch Golden Age still-lives not to impart religious teaching through symbolism but to make us reconsider our relationship with plants and nature from multiple perspectives. At times he represents weeds as metaphors of the LGBTQA+ community. His lush and extremely detailed pastels and drawings cast the human body as a queer reconfiguring that stems from personal mythologies. In this way, he crumbles the nature/culture dichotomy, exploring at once the construction of nature as a concept and how its normativity impacts identity and conceptions of masculinity.

I think it's important to bear in mind that objectification is always part of our relationship with plants through any media. As I see it, the engagement with plants in art is always a matter of degrees of objectification, not one of avoiding it altogether. Even when live plants are brought to the gallery space and they are allowed to manifest some kind of agency—like in the many instances of artists who wire plants to synthesizer or other sound equipment—the plant's status as a subject remains very questionable. Animal studies ended up stuck in a philosophical dead-end because of an obsession with a de-objectification of the animal that ultimately led to the acknowledgment that the alterity of the animal is always impossible to fully grasp. So, with plants, I tend to avoid these unnecessary radicalisms that are mostly born of Western philosophical approaches in favor of the perspectives of artists like Rashid Johnson, Jin Lee, Precious Okoyomon, Diana Scherer, and Jenny Kendler who finetune objectification and anthropomorphism in the knowledge that we can never fully escape them but that we can negotiate them so that the resulting work of art presents alternative, coexisting, and interwoven plant-human becomings.

+ Flattening—speaking now in a literal rather than an ontological sense—is inseparable from the desire to classify knowledge through standardization in naming, through pressed specimens, and in attempts to calibrate color charts for depicting plant specimens for scientific study. Are there recent works that come to mind that provocatively explore the concept of flattening?

I talk at length about flattening as an essential epistemological step in the history of western natural history in my book, *Speculative Taxidermy*. Ron Broglio proposed a compelling theorization of the "flattening of the animal" in his *Surface Encounters: Thinking with Animals and Art*, that I repurpose and expand in the context of contemporary taxidermy in art. Of all traditional natural history methods, flattening is the most interesting to me because it metaphorically binds taxidermy, herbaria, treatises, and painting/drawing. It's an epistemological, trans-medium process aligned with human thinking – we flatten the world as we turn it into the words that exist upon the flatness of the page. Among the many artists who critically embrace natural history's flattening to problematize its objectifying tendencies is Anaïs Tondeur who collaborated with plant-philosopher Michael Marder on an ongoing series called *Chernobyl Herbarium*. The plates comprised in this anthropogenic herbarium are direct imprints on photosensitive paper of radioactive plant specimens collected within the 30 km exclusion zone around the former

Bioremediation (Kudzu) 2018–2019, by Jenny Kendler. The invasive vine is used to "compost" Confederate monuments.

Chernobyl nuclear facility. Unlike traditional natural history herbaria, which aimed to preserve the aesthetic appearance of plants for taxonomical purposes, Tondeur's images expose conflicted economies of the sublime through which negotiating loss and trauma of colossal magnitude can unfold. *Chernobyl Herbarium* is a silent reminder of the planetary cost that technological advancements entail and that are ultimately incompatible with our safety and, with that, our planet.

+ Those very powerful images from *Chernobyl Herbarium* were featured in a recent issue of this journal (*LA+ GREEN*) to accompany an interview with philosopher Michael Marder, who you are also collaborating with. What do you think artists offer that those who study plant relations from other disciplines such as anthropology, ecology, landscape architecture, or philosophy cannot?

Any being, from the one that appears the simplest to the most complex cannot be adequately represented by the optics of one medium or discipline alone. A discipline is defined by its optic and what it can see through it. That's why I believe in collaboration between the humanities and science. From sexism to racism and the exclusion of Indigenous knowledges (along with many others), the methodological shortcomings of Western philosophy have been exposed. Artists have the privilege of being less constrained by disciplinary methods and optics and as such they can more fluidly operate across disciplines and discourses while forging new methods. Their epistemological approaches can be more playful. They are allowed to fail and build on failure much more readily than other practitioners whose work is constantly stifled by peer reviews and institutional blinkeredness. That's not the same as to say that artists enjoy universal freedom. But their experimental scope is certainly wider. As such they can often point to original and important directions before others can.

+ Thus far we've discussed your work as an author, editor, and curator but you are also an artist working with a number of different mediums and subjects.

I don't think of myself as an artist. I actually find that label very problematic since I believe that it carries a dangerous, elitist meaning that hinders creativity. I have used it as a shorthand in conversations but more to designate a category of creative people working in a certain capitalist framework. I have written a book about this. It will be published this year. I take photographs, paint, draw, and compose and record soundscapes/music. Each medium allows me to think about something from different perspectives – that's the most important thing. I believe in cross pollination among media as an essential part of creative thinking. My writing is where things often come together but that's not to say that writing is always the end of the process. We are all in a way or another trapped by labels that cast expectations in the eyes of others. People tend to think about me as an art historian and thus many expect that the written word is all I should be concerned with. Some artists are very snobbish about this. They assume that art historians don't really understand what being creative is about because they only experience creativity vicariously – that might be true for some but writing can be art and I think about composing music as writing, sometimes, or the other way around.

+ In what ways do plants figure in your own work?

More recently I have been making experimental videos for the tracks featured in my album, *Moths* (Blue Spiral Records, 2022). The tracks capture thoughts and feelings that characterized my experiences as a member of the LGBTQA+ community after I moved to the US, which happened during a time of intense social and political tension. Each track is named after a species of moth that lives where I have lived at some point. Compositional structures and moods are largely defined by the way these crepuscular insects fly, hover through vegetation, spiral around lights, and disappear into darkness. Working on these tracks has over time become a way to process events – an opportunity to collect and reconsider specific moments, fragments, and memories that, like moths, would return to visit at dusk.

While recording the album I began to think about a visual component for each track. I then began to film plants in my garden at night, moving the camera through and around the vegetation in a way that captured a different, more insect-like perspective but with a poetic license. I then slowed down the footage and proceeded to match sections of it to each track. I edit each clip for hours trying to capture assonances between the sound, the melodies, the moods. The result is a sensual encounter with plants that remain constantly elusive and yet present. I experiment with the tension between realism and abstraction—out of focus, oversaturation of color, cropping, etc.– in order to finetune a non-traditional intraspecies sensuality that in a speculative (non-scientific) sense explores elusive, human-nonhuman empathic registers.

+ In your book *Why Look at Plants?* the framing of texts and projects through categories such as forest, garden, greenhouse, store, lab, and so on is particularly effective for highlighting how context structures different relationships not only between plants and their environment but in terms of how the environments themselves are part of larger sociopolitical and cultural networks that give rise to them. Given that the environment within which a plant is encountered affects its reception, do you see inherent limitations in a gallery context in terms of how it bears on the topics we've already discussed, such as objectification and anthropomorphism?

Yes – to bring a plant into the gallery space is an objectifying gesture. The plant does not belong there and it is brought into that space in a way that reduces its enmeshing with the world. In his essay "Why Look at Animals?" John Berger said that zoo animals can only disappoint because they are framed like artworks in a gallery. Extracted from their natural environment, animals turn into living representations of an animality that's being denied to them – they become slivers of their former selves. The same applies to plants. However, there are instances in which isolating the nonhuman in the gallery space can be very productive and draw attention to aspects of their being that would otherwise remain invisible. But we have to be comfortable with the idea that the gallery space always proposes an objectifying paradigm that we can only negotiate.

+ What botanic explorations are next for you?

I have coedited a reader on plants in art and philosophy with Michael Marder. This was published in 2023 by MIT with the title *Vegetal Entwinements in Philosophy and Art*. I have also edited a collection of original texts on Manuela Infante's performance *Estado Vegetal* published by University of Minnesota Press in 2023, and I am currently finishing a book on plants in art for Getty, which is coming out early this year.

MATTHEW GANDY
FORENSIC ECOLOGIES AND THE BOTANICAL CITY

Matthew Gandy is Professor of Geography at the University of Cambridge and an award-winning documentary film maker. His books include *Concrete and Clay: Reworking Nature in New York City* (2002), *Urban Constellations* (2011), *The Acoustic City* (2014), *The Fabric of Space: Water, Modernity, and the Urban Imagination* (2014), *Moth* (2016), *The Botanical City* (2020), and *Natura Urbana: Ecological Constellations in Urban Space* (2022).

+ GEOGRAPHY, BOTANY

What does it mean to regard urban space through a botanical lens, or even a "botanical gaze"? Might this be a portal into a different way of seeing or experiencing urban space? The practice of urban botany holds connotations far beyond the scientific field, ranging from the metaphorical appropriations of the Chicago School to emerging interest in the nonhuman labor of plants within the multi-species city. Clearly, the very idea of urban botany conjures up an intricate field of cultural and ecological intersections, linking specific spaces of nature to an array of global connections and material traces. In my London garden I have been constantly amazed by the strange and often unfamiliar plants that spontaneously colonize any patch of bare ground. A plant called caper spurge (*Euphorbia lathyris*), for example, is of Mediterranean origin, thought to have been introduced for medicinal use in the Roman era, while another species, Canadian fleabane (*Erigeron canadensis*), originating in North and Central America, is simply an adventive species that is indicative of more recent global interactions.[1] If we regard urban plants more closely, and especially the spontaneous flora of marginal spaces, this presents a different kind of analytical vantage point to an animal-oriented conception of urban nature.[2]

In this brief article I want to explore the meaning of forensic ecologies in an urban context, drawing on the insights to be gained from a botanical reading of urban space. I am using the term "forensic ecologies" to denote a conceptual synthesis between the critical paradigm of forensic architecture and the role of indicator species derived from fields such as forensic entomology, forensic archaeology, and other longstanding approaches to the precise reconstruction of past environments.[3] Notably, an emphasis on forensic ecologies can produce counter hegemonic data wherein new knowledge serves to challenge existing foci of power. The discovery of rare plants on marginal urban sites can form the basis of a legally grounded socio-ecological challenge to the speculative dynamics of capitalist urbanization. In the case of Berlin, for example, the sociologist Jens Lachmund's study of the Südgelände former railyards shows how a forensic approach to the collection of data on ruderal flora and fauna was successfully deployed in the 1980s and 1990s to protect a fragment of *Stadtwildnis* (urban wilderness) from development, eventually leading to the creation of an urban nature park.[4] Similarly, in London, the discovery of rare plants in the Walthamstow Marshes such as adder's-tongue (*Ophioglossum vulgatum*) helped to provide the area with a degree of legal protection as a Site of Special Scientific Interest in 1985 after a long-running campaign led by local activists since the 1970s.

The question of rarity is highly context specific. By noticing interesting plants in unexpected places, the urban botanist can illuminate overlooked dimensions of the ecological dynamics of urban space. The history of urban botany can be read through the shifting definitions of weeds and "non-weeds" as different facets of nature become regarded as "plants out of place." Indeed, even the same species can pass through successive stages of cultural symbolism as an object of curiosity or cultivation followed by a phase of neglect or cultural disappearance only to dramatically reappear as an object of contestation and anxiety. A prominent example in European cities is the tree-of-heaven (*Ailanthus altissima*), originating from China and Vietnam, that was first introduced as an ornamental curiosity, then recommended for municipal arboriculture as a drought-resistant street tree in the 1950s and 1960s, but later resurfacing as a "feral" component of urban vegetation in parts of Europe, North America, and elsewhere with dense stands developing along railways, roadsides, and other interstitial spaces.

The compulsion to remove urban weeds can be characterized as a recurring facet of bourgeois environmentalism and the quest for spatial order in the modern metropolis.[5] Consequently, a fascination with weeds forms part of a longstanding counter discourse that we might gather under the umbrella of "observational paradigms" within urban ecology that reaches back to early studies of post-industrial ecologies, ballast floras associated with ports, or the distinctive assemblages of plants associated with old walls and ruins. In Richard Deakin's classic study of the ruins of the Roman Colosseum, for example, published in 1855, over 400 species of plants are recorded from across Europe, North Africa, and further afield, yet the unification of the modern Italian state in the 1870s led to a "clean up" of this and many other archaeological sites in which botanical traces of the past were effectively erased.[6] Urban ruins offer a double aesthetic under modernity: on the one hand we encounter a complexity of form associated with processes of decay and abandonment, which has sometimes been elided with neo-romanticist cultural tropes; and on the other hand, we can observe specific kinds of unusual ecological assemblages, such as those associated with the eroded mortar in old walls, which is often alkaline and can support ferns and other plants associated with biotically rich rock crevices that occur on limestone cliffs or even specialized kinds of alpine habitats.

The idea of ordinary street corners serving as a botanical portal into global history connects with a critical taxonomic reading of urban space. The use of Latin plant names evokes a distinctive kind of abstract scientific cadence for the interpretation of spontaneous vegetation. There is a mysterious quality to scientific nomenclature that has enriched the arts-science interface as part of a variety of cultural and ecological investigations. Since sites of these kinds represent a cosmopolitan urban flora, the Linnaean taxonomic schema can be turned on its head so that the classificatory idioms of modernity can form the basis for a postcolonial reading of urban space. The binomial nomenclature devised by Linnaeus serves as a Janus-faced language of spatial discovery since the precise analysis of actually existing urban nature can dispel nativist conceptions of ecology and landscape. Every empty plot can produce a unique kind of global ecological assemblage through chance combinations of seeds brought by wind, birds, or even human feet. Marginal sites constitute a kind of experimental zone that attests to complex configurations of human and nonhuman agency.

With the extension of phytosociology or "plant sociology" to urban and industrial biotopes in the 20th century we find that Linnaean nomenclature has expanded to encompass specific kinds of plant assemblages associated with marginal spaces. No longer simply referring to a species these additional scientific monikers refer to a series of distinctive plant associations that characterize specific sites or substrates. In the case of marginal spaces in Berlin, for example, we often encounter a plant assemblage named as "Dauco-Melilotion Görs ex Rostański et Gutta, 1971," which is indicative of a characteristic combination of species that can emerge after several years on urban wastelands or *Brachen*. The name denotes a distinctive combination of plant genera derived from the carrot and clover families that has been widely deployed in the scientific appraisal of ruderal vegetation encountered in urban wastelands and similar kinds of sites.[7] An emphasis on putative "plant communities" can also point to alternative conceptualizations of the multi-species city that emphasize rhizomatic or mycelial connections between plants, fungi, and other organisms, and especially the kind of complex associations to be found in soil.

Urban soils represent a kind of living archive. The botanical investigations of the Brazilian artist Maria Thereza Alves have allowed the reconstruction of specific facets of global history such as the global reach of European colonialism and the ecological imprint of transatlantic slave trade routes. In the series of site-specific installations entitled *Seeds of Change* (1999–) Alves examines the flora associated with ballast waste in port cities to reveal how specific species were transported between different locations. The term "ballast" refers to the miscellaneous types of soil, stones, and other materials that ships use to balance out their cargoes but discarded on arrival at their destination. Working in collaboration with botanists, Alves found that soils from these former ballast sites can contain seeds that have remained dormant for decades or even centuries. In a recent iteration of her project *Seeds of Change: Antwerp*, first exhibited in 2019, Alves uses an analysis of ballast flora in Antwerp to uncover links with colonial atrocities committed in Congo, Guatemala, and elsewhere, as well as posing questions about the persistence of nativist doctrines in post-colonial European cities. The complexity of actually existing urban nature unsettles existing conceptions of bioregions or other kinds of bounded ecological phenomena. There is a queering of regional sensibilities that can encompass successive layers of cultural and ecological influences so that existing scientific discourse becomes disoriented in the face of unfamiliar socio-ecological configurations.

Patterns of urban vegetation can both conceal and reveal traumatic events: the scars left by wartime damage, geopolitical division, or economic upheaval can gradually disappear amid a carpet of plants. Over time a "wild urban woodland" can gradually envelop urban and industrial ruins – the roots systematically tearing through concrete surfaces or creeping plants forming a thick blanket of green over remaining structures. An example of this process is the gradual transformation of the wartime rubble mountain known as the Teufelsberg (Devil's mountain) in former West Berlin comprised from the debris produced by the aerial bombardment of the city in the closing stages of the Second World War.[8] In the 1970s this stony wilderness harbored many plants adjusted to warm and dry environments such as sticky-leaved goosefoot (*Dysphania botrys*) of Mediterranean origin and various species of tumbleweed (e.g., *Salsola* spp.) from the eroded landscapes of the American Midwest. In recent years, however, a dense forest of adventitious trees such as sycamore (*Acer pseudoplatanus*) has given the landscape a verdant appearance, an ecological trompe l'oeil, in which the recent hillside forest appears to connect with the ancient Grunewald forest toward the edge of the city. At ground level, however, as we enter the shaded interior of this urban forest we encounter the fragments of lost lives: small shards of tiles and other ceramic items that were once part of people's homes litter the surface of the ground in all directions.

FORENSIC ECOLOGIES AND THE BOTANICAL CITY

The practice of urban botany is closely enmeshed with walking methodologies. Walking itineraries provide a multisensory immersion in urban space in which plants can provide rich insights into ecology, memory, and multi-scalar connections over time, connecting urban ecological discourse with histories of modernity, colonialism, and disparate cultures of nature. A botanical itinerary through urban space can be characterized by a certain gait or pace, as the walker becomes immersed in a kind of ecological reverie. There are sensory filters in play here too: as some things come into view others become invisible. Signs of life springing from the sidewalk become an intricate arrangement of lifeforms just as the hum of traffic recedes from conscious awareness. The texture or smell of leaves held in the hand can become an ecological microcosm in thrall to all the senses.

Botanical forays through urban space can constitute an "ecological ethnography" that is derived from slow and often repeated itineraries through different parts of the city. Marginal or overlooked spaces are transformed into sites of insight and discovery, even to the point of unsettling existing approaches within the ecological sciences. Indeed, an expanded conception of botany is an inherently interdisciplinary field that underlines an enduring arts-science interface associated with the meticulous observation of ordinary spaces of nature. Botany provides a distinctive way of experiencing urban space that extends to all dimensions of the human sensorium and beyond. By extending the practice of urban botany to the wider field of post-positivist forensic ecologies we can explore multi-scalar dimensions of urban space that connect with multiple human and nonhuman temporalities. The production of counter hegemonic forms of botanical knowledge can protect specific sites, reveal hidden histories, and also point to more complex configurations of agency.

1 See, for example, Robert E. Witcher, "On Rome's Ecological Contribution to British Flora and Fauna: Landscape, Legacy and Identity," *Landscape History* 34, no. 2 (2013): 5–26.

2 See Marion Ernwein, "Bringing Urban Parks to Life: The More-Than-Human Politics of Urban Ecological Work," *Annals of the American Association of Geographers* 111, no. 2 (2021): 559–76.

3 See Matthew Gandy, *Natura Urbana: Ecological Constellations in Urban Space* (MIT Press, 2022). For an introduction to the work of forensic architecture see Eyal Weizman, *Forensic Architecture: Violence at the Threshold of Detectability* (Zone Books, 2017).

4 Jens Lachmund, *Greening Berlin: The Co-Production of Science, Politics, and Urban Nature* (MIT Press, 2013). See also Ingo Kowarik & Andreas Langer, "Natur-Park Südgelände: Linking Conservation and Recreation in an Abandoned Railyard in Berlin," in Ingo Kowarik & Stefan Körner (eds) *Wild Urban Woodlands: New Perspectives for Urban Forestry* (Springer, 2005), 287–99.

5 Zachary J. S. Falck, *Weeds: An Environmental History of Metropolitan America* (University of Pittsburgh Press, 2010).

6 Richard Deakin, *Flora of the Colosseum of Rome; or Illustrations and Descriptions of Four Hundred and Twenty Plants Growing Spontaneously upon the Ruins of the Colosseum of Rome* (Groombridge, 1855).

7 Matthew Gandy, "Ghosts and Monsters: Reconstructing Nature on the Site of the Berlin Wall," *Transactions of the Institute of British Geographers* 47, no. 4 (2022): 1120–36.

8 See Volkmar Fichtner, *Die anthropogen bedingte Umwandlung des Reliefs durch Trümmeraufschüttungen in Berlin (West) seit 1945* (PhD dissertation, Free University, Berlin, 1977).

Opposite: Illustration of ruin flora in Richard Deakin's *Flora of the Colosseum of Rome* published in 1855.

BOTANIC LESSONS

BERONDA L. MONTGOMERY

Beronda L. Montgomery is a professor of biology and Vice President for Academic Affairs at Grinnell College, Iowa. Her articles on plants and the lessons on living and thriving that we can draw from them have appeared in *American Scientist, Nature,* and *Elle,* as well as in a column for *New Scientist.* Montgomery wrote the book *Lessons from Plants* (2021), discussed on numerous podcasts including *Getting Curious* with Jonathan Van Ness, *Science Clear + Vivid* with Alan Alda, and NPR's *1A*.

+ BOTANY, HEALTH

FROM THE PRAIRIE

P lants often fade into the background of our day-to-day existence. They can provide a largely overlooked yet beautiful green backdrop or an aromatic touchstone. Indeed, if you've walked by blooming honeysuckle or rose bushes on a warm spring or summer day, the scents can stick with you and serve as a trigger for distinct place-based memories for years to come. I can distinctly remember long summer days filled with play, laughter, and honeysuckle nectar snacks in the yard of my childhood home. Anytime I encounter honeysuckle essence, I am immediately transported back to joy and tranquility of those leisurely days with siblings and friends.

Plant scents have been shown to trigger memories, positively impact human mood, and have been implicated in improvements in both human physical and cognitive states.[1] Close observation of plants also has the potential to provide what I think of as principles of the universe on thriving and persistence, much like observation of nature more broadly.[2] I think of these as "lessons from plants" that can serve as examples and inspiration for human thriving.[3] As a plant scientist and general botanic enthusiast, I look for these lessons everywhere. I have recently been deeply impacted by botanic lessons on recognizing and responding to stress.

Last summer, I moved to a new academic campus. One of my first activities when I arrive in a new location is to get to know my new plant cohabitants. Having moved to Iowa in the Midwest United States, I'd taken up residence in a location with much to offer in terms of new plant neighbors. The campus is in the core of agricultural country. The primary botanical crop of the region is maize or corn and you can drive through literal miles of cornfields as you traverse the region. Yet historically the area was largely tall grass prairie where milkweed, purple coneflower, and cord grass were common plants.[4]

I am fortunate to live and work in an area that has restored and preserved prairie habitats. There is a prairie community that I pass daily on my way to work. In my first year in this new location, I arrived mid-summer when the prairie patch was in its full glory. There were mature milkweed plants visited daily by beautiful monarch butterflies, tall pale purple coneflowers catching the sunlight and providing optical, and artistic, contrast to the yellow milkweed flowers, and bright green grasses that in addition to providing attractive greenery function to slow water runoff during rain, which improves local water quality.[5] Beyond being linked to such ecological benefits, prairies provide habitat for a range of wildlife, big and small, including pollinators and other insects, birds, amphibians, and reptiles.[6] As humans are often more apt to acknowledge animals over flora, we may be more likely to lament the loss of the wildlife than the plants when a prairie recedes.[7] Some of us may be more likely at least. I am finely attuned to my plant neighbors and, noticing the failure to thrive of several plants in the local prairie community, I immediately began to ask questions about what was causing the yellowing of the leaves of some plants and apparent wilting or atrophy of stems of others. Yellowing of leaves is technically due to a chlorophyll deficiency; yet, yellowing can indicate a number of common causal factors – nutrient deficiency including, commonly, iron or nitrogen deficiencies; pH imbalance in soils causing alkalinity or acidity; crowding of plants in a particular area; over- or underwatering; or, senescence or aging, among other factors.[8] Of note, yellowing of leaves paired with deformation of leaf shape is much more likely to indicate bacterial or fungal infestation.

When noticing yellowing of plant leaves, good plant caretakers seek to identify the environmental factors that may be contributing to it and to mediate them to promote recovery of greening and health in plants. Doing so successfully requires careful observation to identify the stress that results in the plant's yellowing, an ability to identify potential causes and effective interventions to reverse them, and consistent care to monitor recovery from stress. My recent observations of select yellowing of leaves of plants in the campus prairie caused me to reflect on what we do in a campus community or social community when we have individuals in our midst exhibiting stress. Whether it is work-related stress, public health-related stress such as that observed in the ongoing global pandemic, or economic stress that can result in food or housing insecurities, what is our response, and does it mirror the responses we have when we observe plants exhibiting stress?

All too frequently our response to individuals in our human communities exhibiting challenges or stress is to default to assessment of personal deficits to thrive.[9] It is not uncommon to attribute such challenges to poor personal choices and a lack of personal responsibility.[10] However, our default interactions with plants, and often animals, is to expect thriving and success in natural ecosystems and to assess departures from the expectations as failures in access to critical factors in the ecosystem of residence, rather than individual organismal failures to access and use them.

Every time I encounter a yellowing plant, I ask what does it need in order to thrive? Plants in obvious distress remind me of how our human communities would be transformed—for the better, I believe—if we engaged with our fellow humans with a parallel expectation of growth and thriving. Where would it lead us if we asked what is missing in the ecosystem of residence rather than ask what is wrong with our fellow humans?

After reporting the observed stress to and consulting with local campus horticulturalists, I watched carefully to see if my plant neighbors would recover. As most of them slowly returned to the green color that represents health and vitality in plants, I silently (and occasionally very vocally) cheered them on. In the instance where one or two were unable to rebound despite the most well-intentioned interventions of their devoted caretakers, we collectively felt bad that we had not been able to support them in their recovery from stress. My greatest honor is to endeavor to apply this commitment to recognize stress, intervene, and hope for the best equally with the plants, humans, and other organisms with whom I am privileged to cohabitate.

1 Eran Pichersky, "Plant Scents: What we perceive as a fragment perfume is actually a sophisticated tool used by plants to entice pollinators, discourage microbes and fend off predators," *American Scientist* 92 (2004): 514–21; S. Jiang, et. al., "Effect of Fragrant Primula Flowers on Physiology and Psychology in Female College Students: An Empirical Study," *Frontiers in Psychology* 12 (2021): 607876.

2 I have shared frequently on such lessons from nature, particularly lessons from the plants and microbes that I study. I have written broadly on lessons from plants and microbes: see, e.g., Beronda L. Montgomery, *Lessons from Plants* (Harvard University Press, 2021), and Beronda L. Montgomery, "Lessons from Microbes: What Can We Learn about Equity from Unculturable Bacteria?" *mSphere* (October 28, 2020).

3 Montgomery, *Lessons from Plants*.

4 See Iowa Department of Natural Resources, https://www.iowadnr.gov/conservation/prairie-resource-center (accessed December 21, 2022).

5 Ann Perry, "Prairie Restoration Also Helps Restore Water Quality," *USDA* (February 28, 2012).

6 Katy Heggen, "Remnant Prairie: A closer look at Iowa's rarest landscape," *Iowa Natural Heritage Foundation* (August 24, 2017).

7 Sandra Knapp, "Are humans really blind to plants?" *Plants, People, Planet* 1 (2019): 164–68.

8 Arit Efretuei, "When Chlorosis is Caused by Nitrogen Deficiency," *Permaculture News* (November 16, 2016); Rebecca Therby-Vale, et al. "Mineral Nutrient Signaling Controls Photosynthesis: Focus on Iron Deficiency-induced Chlorosis," *Trends in Plant Science* 27 (2022): 502–509.

9 Beronda L. Montgomery, "Planting Equity: Using What We Know to Cultivate Growth as a Plant Biology Community," *The Plant Cell* 32 (2020): 3372–75.

10 Matthew Anderson, et. al., "Homelessness and the American dream: An Inconvenient Truth," *Research Outreach* (May 17, 2021).

Jared Farmer is the Walter H. Annenberg Professor of History at the University of Pennsylvania. His research focuses on environmental history in the long 19th century, specializing in the North American West. He has written numerous essays on human relationships with plants and is author of the books *On Zion's Mount: Mormons, Indians, and the American Landscape* (2008) and *Trees in Paradise: A California History* (2013). **Karen M'Closkey** spoke with Farmer about his latest book, *Elderflora: A Modern History of Ancient Trees* (2022), and the lessons that some of the world's longest living organisms offer us in the face of rapid climate change.

+ You have described your work as "place-based planetary history," which sounds contradictory. Can you explain what that means and how trees were a medium through which to approach history in this way?

It's meant to sound paradoxical. No single person can possibly "do" planetary history, but I wanted to write a book about the most important issue of our time, which is climate change, and the scale of that is in fact the planet. Planetary history is different than "global history" and "world history," terms that have been around for decades. World history generally means large-scale geopolitical history, while global history covers processes of globalization, including geopolitics, though these two approaches have somewhat different theoretical bases. Global history is weighted toward the modern period—the consequences of the Columbian Exchange, the Atlantic slave trade, Western imperialism, settler colonialism, capitalism, industrialism, and so on. Planetary history encompasses all of that, and more. It is more than human. It foregrounds nonhuman things because people are now planetary agents who are altering the biosphere and evolution itself—and yet they're not the only changemakers at that scale. They're actually late to the game. Plants have been changing the biosphere a lot longer than humans. So, the story must start millions of years before the modern period. Even though climate change has accelerated in recent centuries and decades, the planet has been through many different climate regimes in its deep history.

I do think trees are one of the best ways to get at planetary history. Woody plants, especially gymnosperms like conifers, come from incredibly old genetic lineages. They've lived through many planets, one place at a time. Each organism is rooted in place for the entirety of its life. Trees are the most place-based of all organisms because their commitment to the local is absolute. And yet they live almost their entire lives outside the favorable climate conditions of their germination. Trees are experts in living out of time while showing resilience in place. They're useful to think with by analogy as we confront climate change. We all now live outside our birth climate, even if we're young—maybe especially if we're young.

+ Dendrochronology—the science of reading tree rings—is so familiar from grade-school science class, it's easy to forget how remarkable it is that humans figured out we can read climate this way. Has the science of dendrochronology changed much in the last few decades, or have there been technological advances that have changed what we thought we knew?

A lot of people assume that tree-ring science must be easy because it means counting rings; they're surprised to learn it was institutionalized so late, in the 20th century. It's actually a very technical and difficult interdisciplinary field concerned with the interpretation of data in cambium layers after they've been dated.

Going back thousands of years, people noticed that woody plants produce growth layers. The notion that these could be annual markers is also pretty old. Leonardo da Vinci was one of the first people to write about it; there were some German naturalists in the 19th century who thought about whether cambium layers in the temperate zone could be read as records of annual weather, and Henry David

Thoreau was fascinated by tree-rings and journaled about them. But it wasn't until an early 20th-century astronomer, Andrew Ellicott Douglass, began thinking about the possible biological effects of the sunspot cycle that dendrochronology came into being. Douglass's technique involved much more than counting rings. It was about connecting rings from different trees and creating networks of samples that shared common signals that could be used as proxy data for environmental phenomena. You need the right trees in the right places to create a cross-dated network; and to pull out all your core samples, you need an increment borer, a tool that wasn't invented until the 19th century.

Certain conifers living in certain ecotones produce the most sensitive rings, and therefore the most reliable data with the least noise. It turns out that the American West—Douglass worked at the University of Arizona—was ideal for tree-ring science because it's a land of mountain conifers. Here Douglass and his protégé Edmund Schulman found members of the cypress and pine families growing in rugged, semi-arid habitats, living in feast-or-famine conditions, depending on whether it rained each summer, and their tree-rings responded in an exaggerated way to precipitation or temperature signals. Within several square miles of steep terrain, they might locate discrete populations that allowed them to get corresponding data samples going back thousands of years. Though Douglass was never able to prove that the sunspot cycle had an effect on tree growth, his cross-dating technique became indispensable to archaeology (for dating architectural structures) and later to climatology.

+ In your book *Elderflora*, you reference Pliny's *Natural History* (77 CE), which you describe as the "oldest extant list of oldest trees in the world." Does this mean that the identification of elderflora is not a recent scientific quest? Or how have the reasons for this ambition to identify elderflora changed?

It's an ancient quest that changed in the modern period. Pliny and other naturalists in the prescientific era emphasized what I call "relational age"; that is, the oldest tree was *as old as* a city, or *as old as* a temple. It wasn't necessary to know exactly what year the tree germinated, or exactly what year the temple or the city was built. Rather, the relationship between the tree and the built environment was the important thing. A consecrated tree could die and be replaced with an offshoot, and effectively live on as the same plant. The "tree" was in fact a *relationship* between people, a cultural tradition, a built landscape, and a vegetal germplasm. This relationship in place embodied cyclical time. Fast forward to the 18th century, and the whole epistemology begins to change. Now, scientists and foresters want to know *exactly* how old an individual tree is based on its cambium layers. Each tree-ring has an exact date or no date. I call this "cambial age." Because of dendrochronology, we now know the precise ages and locations of hundreds of millennial trees – this despite the incredible deforestation of the industrial period, often in combination with settler-colonial land clearance. These surviving plants are irreplaceable by nature: once they die, they are dead, and cannot be replanted. They represent linear time and the terminal nature of modernity. Many of them have become de facto sacred sites. One of the main themes of my book is this recapitulation of modern science and premodern religion. I'm struck by the correspondence between the consecrated trees of the ancient world—sacred because of their purported age in relation to human landscapes—and these more recent "secular sacred trees," which are sanctified by science and guarded not by priests but by park rangers. They're protected because technicians have determined they are old beyond a certain number of years—1,000—significant to rationalists and religionists alike.

On the one hand, I'm moved by, and I celebrate, continuity in human history across periods, across cultures, as exemplified by this desire to identify and venerate slow plants. On the other hand, I find something disturbing in the modern obsession with quantifying the oldest, the biggest, the thickest, the tallest. All those 19th-century naturalists traipsing the globe with measuring tapes can seem silly, but my discomfort runs much deeper. It has to do with our numerically obsessed culture and its privileging of things that can and must be counted and datafied, faster and faster. Science is now one of the leading forms of meaning in the world. In this worldview, if something—including a tree—cannot be datified, does it have no value?

+ You've described trees as "living bridges" between past, present, and future, but have commented that humans have a great ability to think of the deep past but are bad at thinking of our long-term future. Obviously, trees help us understand the past—we've discussed dendrochronology—but how can they help us think about the deep future beyond their role as carbon sinks?

In the process of writing my book, I grew really weary of people saying, "We just need to plant more trees," which strikes me as ludicrously wishful thinking and implicitly offensive, as if woody plants are nothing more than carbon sequestration devices for our instrumental use.

I also got tired of hearing this: "We just need more climate data and then people will finally act." Lack of data is not the problem! It's a collective action problem. It's a political problem. It's an ethical problem. The multi-disciplinary science behind climate modeling was one of the great accomplishments of 20th-century science, but, at this point, politically speaking, we don't need any more data. Tree-ring archives are melancholy spaces—though they do smell heavenly—because the past recorded in the wood samples is no longer a reliable guide to the future. When these collections were established in the 20th century, the idea was to use tree rings to predict climate cycles. That's now impossible. You can still do amazing things with tree rings, as shown by the recent dating of stellar proton events and supervolcanic eruptions, but the climate of these data-recording trees is obsolete.

Likewise, it's undeniable that all the oldest surviving trees of the Holocene will die before their time. They will perish in cohorts and they cannot be replaced, even with cloning, because they have cambial age rather than relational age – though I suppose you could say their mortality is relational to the Anthropocene.

Pronouncing that ancient trees are going to die prematurely is not the same as predicting that tree species will go extinct. Some long-lived species, especially in the conifer division, are in big trouble; others aren't. But there will be a loss of oldness as measured by science. I anticipate an interregnum when most people on the planet will not have easy access to trees absolutely known to be 1,000 years old or older. That's something—an extinction of experience—to mourn. But that's different from saying that there won't be, or can't be, old trees in the future. We need to care for the future oldest trees; we need to design for trees with a future. That's why I'm interested in premodern, prescientific traditions of venerating old trees. I think we need to go back to practices of venerating old things in a relational sense, within built environments. When thinking about the deep future, it often helps to consider the ancient past.

+ Does this relate to your idea of "chronodiversity?" Can you explain the importance of its meaning and how it relates to trees in particular? Does it have a scientific basis or ethic, like biodiversity?

Chronodiversity is complementary to biodiversity. An ecosystem that has species and populations of different evolutionary and organismal ages is going to be a more bio-rich environment. I think of chronodiversity as "temporal richness," an analogue to species richness. If scientists were to take up my vocabulary, they would want to measure it, but in my mind the gifts of elderflora include hard-to-quantify "temporal services," like the experiential, ethical, and spiritual values of being in close contact with an organism that experiences time very differently than you or I do.

Old trees provide emotional access to timefulness. This experience is vital, I think, and people should have access to it. But the experience shouldn't require a wilderness reserve: not everybody will be able to visit Sequoia National Park. This is where design comes in. A great city is one that has a range of old things and beings from different periods, preserved together through use, maintenance, and care. So, in the context of design, the problem of timefulness includes an acknowledgment that the mature

trees of today's urban canopy are probably not the appropriate trees for the future. Can we redesign our cities for a warmer future that will still support megaflora and elderflora? Can planners and designers imagine cities adaptable and perdurable enough to withstand the period of overshoot we're entering? Landscape architects and arborists need to work together now so that we have old trees fifty years from now, and so that we have a continuity of oldness in the midst of novel change and destabilizing loss. I'm comforted by the idea that the oldest trees of the future are here already, in need of our care.

+ In the epilogue to *Elderflora*, you relate the story about how the oldest living thing ever known was killed in the act of knowing—an almost 5,000-year-old bristlecone pine called WPN-114 was cut down in 1964 in what later became Great Basin National Park, Nevada. When hearing a story like this, we think, "That would never happen today." Yet, as you've said, the drive to know, and to quantify, seems only to be increasing. However, I also wonder if such small, shriveled trees, which are elderflora but not megaflora, don't inspire the same veneration as a giant sequoia.

One of the narrative lines in my book is how—because of tree-ring science and its emphasis on quantitative data—a whole class of smallish, knobby, gnarly trees could suddenly become technically valuable, and close behind that, culturally valuable, even worthy of veneration. It's fascinating that people could look at a Great Basin bristlecone pine in the 1940s and see it as kind of ugly, appropriate for fencing or firewood, and then, in the 1960s, see it as sacred. WPN-114 just missed the cut off, so to speak. John Muir visited the Nevada mountains in the 19th century and didn't think bristlecones were anything special. Neither did US foresters in the early 20th century. From their point of view, these pines weren't proper forestry trees. The White Mountains of California/Nevada, where the oldest populations grow, doesn't look like a forest; it's part of Inyo National Forest only because Teddy Roosevelt signed an executive order to protect the watershed for Los Angeles. This presidential move was aligned with L.A.'s water grab – the Owens Valley aqueduct.

My starting position was simple: The more that people care about little gnarly trees, the better. I'm all for enlarging the sphere of ethical concern for woody plants, including shrubs. But then, at some point in my project, I developed some discomfort when I recognized that dendrochronology has a bias toward certain plants in certain areas. There's a built-in bias toward the temperate zone because tropical growth layers do not follow a regular annual pattern. And dendro researchers exclude plants that don't produce cambium, even though succulents can be long-living, too. Tree-ring scientists need plants that generate annual rings that are sensitive to climate signals—or some other kind of signal they're interested in—and produce resinous wood that doesn't decay quickly or hollow out over time so they can capture complete records from the trunk. These data demands rule out most plants, even most woody plants. There are

plenty of big old trees in the tropical zone that, until recently, dendrochronologists ignored. Their discipline was built on the principle that arboreal data must be dated exactly to the year. If you can't put a date on it, it doesn't exist from their point of view. That troubles me. As much as I admire dendro people and write in defense of instrumentally identified elderflora, I believe that numbers are hardly the only measure of oldness, nor the only measure of age-value.

+ You write about what contributes to an individual plant's longevity beyond its natural ability for a long life. You call these factors "placeways." One of them, not surprisingly, is remoteness from humans—protection through isolation—and those adapted to marginal habitats—hot and dry, or steep and exposed, or nutrient poor. However, you also cite many examples where some of the oldest living things are found in temples, churchyards, and shrines. Do you think this type of small-scale, more individual care has been as important culturally—in terms of reaching a broad public about the value of trees—as has large-scale conservation through federal and state protection?

In the modern period, there's a kind of land-use bifurcation. Instead of a variety of premodern woodlands inhabited by people, you get largely unpeopled zones with or without trees: forest preserves, tree-crop plantations, deforested pasturelands. The same nations and empires that are busy deforesting on an industrial scale—often accompanied by the violent displacement of Indigenous peoples—are establishing national parks and forest reserves managed by seasonal rangers and technicians. So, today, you tend to find the oldest populations of trees in remote protected areas with few or no fulltime residents, with the qualification that many of these locations did not always lack people.

But yes, many individual elderflora are found not in national parks or forests, but in churchyards, shrines, temples, municipal parks, state cemeteries, royal compounds—centrally located sites of cultural significance. It's a really old landscape tradition—not quite universal, but incredibly widespread. You find it in many different cultural hearths: East Asia, South Asia, Africa, Mexico, the Mediterranean. Some of these old-tree traditions have died out, but a lot of them, remarkably, have continued to this day. Sometimes they've continued despite the demise of associated empires or religions. If you're a tree and you want to be really old—here I'm anthropomorphizing, of course—you want to be as close as possible to tree keepers. You want to affiliate with humans who come from cultures with deep traditions of caring for trees. You want to be in their center place. That's one kind of placeway. In the modern period, the other main placeway, as I mentioned, is far away as possible from people, and especially certain types of people from the Global North—settlers, capitalists, industrialists—who want to kill you. So, the old-growth habitats of last resort end up being desert mountains, lava fields, swamps, and so on. In the US West, many areas that were never homesteaded or otherwise claimed ended up being included in national parks or forests because they were "useless" in the historic sense of how the US settler state defined use value.

The history of California shows how Americans were tree killers before they were tree keepers. It was not a foregone conclusion that giant sequoia would be preserved. Logging companies clearcut, and dynamited, one of the greatest groves on the planet before the US state decided to start reserving Sierra forests for watershed protection and for scenic enjoyment. Before that, Americans invented entertaining ways to humiliate sequoias by reducing them to stumps, dancing on them, putting bowling alleys on them, using their flayed bark to create parlor rooms and traveling exhibits for P.T. Barnum in New York City and the Crystal Palace in London. In the end, the giant sequoia was just grand enough and rare enough to be protected as the first national park dedicated to plants. It's incredible to think that coast redwood—which historically covered two million acres in California—was not considered sufficiently grand until almost all the old growth had been turned to shingles and sawdust.

+ I have a question about *ex situ*, rather than *in situ*, conservation. Botanic gardens are key in terms of the care and cultivation of plants. But because they are gardens and not wild, and they contain many non-native species, some would argue they cannot contribute significantly to biodiversity conservation, even though many botanic gardens have shifted their missions to do so. Given your research into elderflora, would you say these institutions are significant for conservation through education given the numbers or people who can visit them?

I would say yes, because conservation is more efficacious in an arboretum than a zoo. A garden designer can make a landscape that is beautiful, educational, and spiritually uplifting as well, but that also has a conservation mission. You can "backup" a plant, so to speak, much more easily than an animal. Although one small botanic garden can't by itself keep a species going, if various gardens in similar climate zones maintain as living specimens enough genetic diversity of a species, that's a pretty good backup system because trees produce so many seeds over their long lives. There are, of course, seed banks, too—Kew Gardens at Wakehurst has the largest frozen tree-seed collection in the world—but just the living collection, if it's well tended, can persist on decadal or centurial timescales. That's a backup in case wild populations need assistance if they encounter a genetic bottleneck.

One of my favorite examples is the Royal Botanic Garden Edinburgh. It maintains samples of *Taxus baccata* (common yew, European yew, or English yew) in hedge form. Yews grow compactly and are very long-lived. The arboretum in the Scottish capital has a single hedge with genetic material from hundreds of different yews. They have samples from Great Britain, from Europe, and from remote sites in the Atlas Mountains of North Africa. It's a genetic bank for the species, and it's also an homage to the sometimes reviled but quite wonderful British landscape tradition of yew topiary. To me, that hedge is a great example of conservation meets design meets education.

+ Many botanic gardens also have bonsai collections. Some bonsai are estimated to be about one thousand years old. One could see bonsai as the epitome of anthropocentrism—molding living things to human desires and keeping them contained in highly controlled environments—but they are also a perfect example of thinking in tree time, of practicing, to borrow your phrase, "obligations across generations." Might they be our primary access to elderflora in the future?

The Japanese arborists who developed the technique of tree miniaturization created an aesthetics of longevity. Edmund Schulman, a great tree-ring scientist I profile in my book, developed the maxim "longevity under adversity." He found the world's oldest trees not in clement conditions but in harsh habitats. And his high-altitude pines appeared a bit like bonsai. If you go to an arboretum and look at a bonsai collection, you'll see these plants are put under incredible stress. Even though they are continuously cared for, they're being manipulated and intensively stressed to produce sculptural forms that are reminiscent of naturally long-lived, multi-millennial conifers, even though some of these potted plants are quite young.

Although I'm not moved by bonsai as much as maybe I should be, I do love the practice as a metaphor. I love the idea of an organism passed down in one family for generations, and the obligations that come with that. The obligations are between human and plant, and even more between generations of human kin to care for that plant on a near-daily basis to keep the tradition going. There's beauty in that.

+ With climate change, there are points at which ecosystems that have persisted for thousands, or tens of thousands, or, in the case of rainforests, many millions of years, transition to an alternate state—what is called "vegetation type conversion." You write about this in your book, noting that we are heading toward the "age of the mediocre forest." There is a bit of humor in that phrasing, though perhaps not when we consider the reasons we're headed this way. What does this portend for elderflora?

I borrowed that funny phrase from a scientific article.[1] I'm trying my best to embrace the word "mediocre" because in our climate-changed future, when the weather will only get weirder—instead of heat, it'll be a scorcher; instead of wind, it'll be a derecho; instead of rain, it'll be a flood—there's going to be a lot more damage to trees. Trees can take a lot, but because they can't move, they can't avoid injury when it comes. So, except in highly manicured environments like Penn's campus where there may be arboricultural methods—and budget lines—to prepare for these events, a lot of trees are going to be under big stress time and again. Sure, there'll be plant migrations—some assisted by humans, others happening on their own, as has always been the case. But even those trees that thrive in a changed climate with more intense disturbances will look different because they'll be subject to more stress and injury. They will respond by growing differently. They might put more energy into stability than verticality and crown growth.

I'm not worried about the death of the forest per se. Although some trees and species won't make it, we're in no danger of becoming a treeless planet. Future forests may be less biodiverse, and tree cover less photogenic, but nonetheless I want to imagine a world that is hotter and weirder yet still full of woody plants we care for – even though those cared-for plants look different. We might need to develop a new aesthetic to appreciate the forms that trees will take. Perhaps the bonsai tradition could be a model for seeing injuries as marks of history that have their own beauty. We would benefit from an aesthetic that sees "mediocrity" not as loss but as adaptation or perdurance, meaning resilience over time. Trees have lived through so many different climates on the planet going back hundreds of millions of years; given enough space to find their place, they'll make the adjustments again. It will be painful for us to see so many big old trees pass away. But keeping our relationship going with elderflora, even as the trees themselves change in form and population – that's what matters. It's our long-term relationships with long-lived things that matter most.

---

1 Steven G. McNulty, et al., "The Rise of the Mediocre Forest: Why Chronically Stressed Trees May Better Survive Extreme Episodic Climate Variability," *New Forests* 45 (2014): 403–15.

Patrick Blanc is a French botanist, researcher, and inventor of a simple technique for growing plants on vertical surfaces with no soil and minimal maintenance. A specialist in rainforest understory plants, he has created hundreds of indoor and outdoor vertical gardens in a wide range of environments globally. Catherine Mosbach is a Paris-based landscape architect. Founded in 1987, Mosbach Paysagiste has designed many award-winning projects including the Bordeaux Botanic Garden, The Louvre-Lens Museum grounds, and Taichung Central Park. Blanc and Mosbach spoke with **Karen M'Closkey** about their collaboration, the differences between working as a botanist and a landscape architect, and the relationship between design precision and the spontaneity of life.

+ Catherine you are a landscape architect with a background in the life sciences, and Patrick you are a botanist who has traveled the world to study plants, in addition to creating hundreds of your own designs. What projects have you worked on together and how did your collaboration begin?

CM We met for the publication *Pages Paysages*, which I founded in 1987 with Marc Claramunt, Pascale Jacotot, and Vincent Tricaud. The journal was a wonderful opportunity to merge scientists, philosophers, artists, and other practitioners. We wanted it to be about experimentation and exploration of different design approaches. I invited Patrick to write a paper about the work he had done on vertical gardens for the garden festival Chaumont sur Loire 1998. The title of the text we proposed to Patrick was "Be a Plant."[1] It was a powerful paper, very philosophical. Then we (Mosbach Paysagiste) won the Bordeaux Botanical Garden Competition in 1999, which was one of our first major commissions. In the past it was normal practice for botanic gardens associated with universities to be experimental and so it was absolutely necessary for me to not simply make a public space but to make a space that was experimental in terms of how it represented landscapes of the Aquitaine Basin region. Patrick worked with us on the tree species and climbing and aquatic plants in a didactic way. It was a great collaboration.

+ The plants available from nurseries are a tiny fraction of the over 400,000 known plant species. Can you speak about how you source plants and if you've seen changes in the horticultural industry over the last few decades?

PB I am lucky to be able to travel all over the world. In some countries I can obtain local plants for my projects, while in other countries, it's almost impossible to have local plants. I first prepare a list of the plants I would like to use and then make some modifications according to what is available from local nurseries. But plants from local nurseries does not necessarily mean native plants. In Europe, nurseries have plants that come from many places – some from Europe, many from Malaysia, Japan, China, very few, unfortunately, from the United States. This is very different from, for instance, California where many nurseries specialize in native Californian plants. The same is true in Australia, New Zealand, and South Africa. But in Europe, or in other areas in the US outside of California, unfortunately, it is not possible. It's a pity because when I visit the US and see all of the interesting species in the Appalachian Mountains, in all the states south of New York to Florida, all of North Carolina, South Carolina, Virginia, I wonder, "Why don't they cultivate these species?" For some projects, that is what I do when the plants are not available. In Japan, for a project for the Shinkansen, I observed that the mountains surrounding the city of Yamaguchi were covered with forest. Because we had two years before the project had to be realized, I proposed that we go to the forest to gather some cuttings or seeds to propagate. So, it was wonderful because we–the architects and people from eight nurseries–went into the field. The nurseries then propagated the plants for two years – all 150 species used in the project come from a 15 km area around the Shinkansen Station.

CM In Europe, it's quite easy to get species from all over Europe. It's not protected, it's open. The nurseries have to be specifically managed, however, so there is a high standard in terms of the plant stock. But in a place like Taiwan, for our Taichung Central Park project, it's totally different because it is an island and you can only use what is available on the island so as not to bring pathogens or possibly invasive plants. At the same time, the natural areas are protected so we could not get native plants from the region. Nor do they have the nursery regulations like we have in Europe – the plants are treated very badly. When they are old, they are respected as gods but the nursery plants are treated like trash so it was very difficult to get the quantity and quality of trees that we needed – 10,000 trees. And nor could we get many of the species we wanted because we could neither import plants nor gather them from the surrounding areas.

+ Do you think projects like the one in Japan where you grew the plants for it can end up changing nursery practices, or do you think it was a one-time situation? And is your preference to use native plants when you can and, if so, what do you mean by that in terms of a plant's geography?

PB It's so difficult to define what "native" is. Native has many different senses, depending on the species. Some species are very narrow organisms, known only from some rocky areas in the world, whereas others have a very wide distribution. Usually, plants growing horizontally in the soil have a quite large distribution area but plants growing on cliffs, for instance along the Mediterranean Sea, have a very narrow distribution. So, I use native plants when there are–like in the San Francisco area–wonderful plants for walls because there are so many rocky areas where plants are adapted to environments with little or no soil. But if you go in the forest of Fontainebleau, I can tell you, there are maybe five interesting species, nothing more. So, of course, it depends. In some cases, it's important to try to use native species, in other cases it's not at all necessary.

But, yes, I do think we can change practices beyond a single project. I am currently working on a project at the Singapore Airport. I brought seeds in my luggage without any declaration, and they were very happy to have new plants to propagate. After two years, there are many new species that they can use in different places beyond the airport.

CM It doesn't happen. I could never have specific production for a specific project. Patrick is a botanist so he has much more precise knowledge to be able to say, "I'm sure it will work" whereas we are landscape architects. Even though we have plant knowledge we are not scientific about it. And we are also open to what emerges in some places – seeds are always traveling across the continent and so there are surprises. It's just a way for a plant to adapt to a new condition, like all organic and life processes.

+ Patrick you spent a lot of time developing the technical assembly and nutrient needs for the vertical garden system but have either of you worked on projects where experimentation is ongoing rather than something that is done prior to a project and then applied to it?

PB My vertical garden structure is very simple. It's just a hose with some holes, on a kind of felt, so technically it's nothing. The only important thing is the selection of the species, especially since the gardens are vertical and, when they're outdoors, you have much more wind and light at the top than at the bottom, and also a big difference between summer and winter in terms of a plant's access to light. It is crucial to select the right species for the right place on the vertical garden. Only then do you focus on creating the best aesthetic value. So this is not experimental. I know the growth habits of the plants in nature, so I try to replicate what they need. It's the same for nutrients. I give nutrients in very low amounts, or in some cases, such as when the garden is fed with water from a fish tank, I do not add nutrients because the fish provide enough. So, again, nothing is experimental. Of course, year after year, I use different plant species and sometimes there are surprises. Some species you see in nature growing vertically on cliffs will not grow well in my vertical gardens but it's not a problem. When you use 200 or 300 species and have one species that is not growing well, others will fill in. Plants are highly adaptable. They come by themselves. This is evolution. This is the miracle of life.

CM When I say experimental, I do not mean in a scientific way, rather, we design some places to be highly controlled, and many other places to be open to the unexpected. But many people are not comfortable with the "unknown," which is a pity. For example, for Taichung Central Park we created mountainous places and planted over 10,000 trees. It was intended to be a forest with its own biological regulation but, in autumn, when the leaves fell on the soil, it was considered dirty and the forest was "cleaned." Of course, this is the opposite of what should happen because the leaves introduce life into the soil. In many places there is a lot of effort to create horticultural displays with plants to make them always colorful and flowerful, which is the opposite of our spirit. So, when I say experimental, it means we design these projects in a way that people can respect and understand that we are pointing to a new direction for landscapes. But it takes a long time, and you need to have the support of those in charge of maintenance who respect the spirit of the project; otherwise, it will be normalized and standardized over time, which cuts out everything spontaneous and takes away what you could learn about creativity and life.

Below: Louvre-Lens prairie planting strategy.

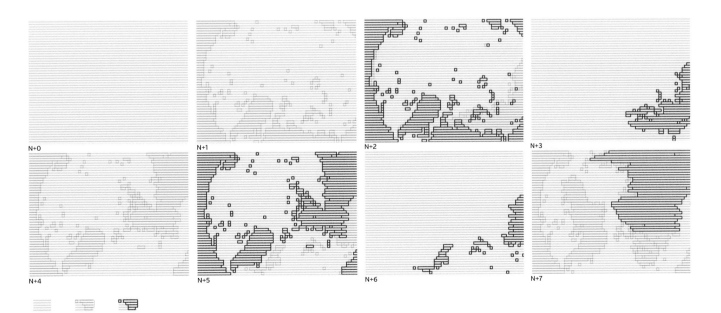

+ How do you engage with those who will take care of these landscapes? How receptive are clients to involving you long term?

CM I am not hired. It's a huge problem. I had a commission, with architect Catherine Frenak, for the Archaeological Park of Solutré (1999–2006) and in that situation they are letting things evolve spontaneously but that is because the soil is protected as an historical monument since it has human remains from the Solutrean (Upper Paleolithic c. 22,000 – c. 17,000 BP), not because of the spirit of our project, which is about sharing knowledge, sharing discovery, and sharing life processes. For example, our design for Musée Parc du Louvre-Lens needs time—especially because the soil is very poor—we need approximately 20 years to see what evolves. It's a much more humanistic way to engage and nourish a dialogue with what you don't know, and accept you don't know, but that depends on the owner of the place. And a huge difference from Patrick is that I work on public spaces. Even if it is a museum or botanic garden, it welcomes the public.

PB It's a little bit different for me because my works—maybe 300 or 400 now—are considered a little bit like artworks. Very often the clients ask me, after some years, about how to maintain the gardens or if they should cut or replace a plant. So, it is very different from working on horizontal space that is accessible by the public and by gardeners. But there are still some problems with the companies selected for maintenance. They sometimes pull-out plants or change them so they can keep coming back for maintenance – sometimes as often as every two weeks. But I have selected my plants carefully and so my gardens do not need very much maintenance, typically only a few times per year.

+ Your first collaboration was the Bordeaux Botanic Garden. Does designing a botanic garden change how you approach planting choice and design as compared to any other landscape project?

CM We won the competition because one of the jurors—botanist Claude Figureau—was a researcher who said ours was the only project that opened the capacity for research. The other entries were static, much more like a museum, and with less interaction between the visitors and the space. I learned a lot about the capacity of bacteria as an input for life in the earth and, had I not done that, I would not have had that knowledge for the Louvre-Lens. I was lucky to work with a unique company, using some techniques that we developed for working with soil parameters and stratigraphic boundaries, and used later for Louvre-Lens. But a space cannot just be for one kind of population; it must be open. The Bordeaux Botanic Garden is quite small—six hectares—but it's really powerful to be in this landscape, in the city center. It works as a public space much better than we expected but is losing some of its links to research. As I said before, it depends on the keeper of the landscape. It is a very delicate connection between the landscape and the gardener or keeper.

+ Speaking of the afterlife of these projects, I'm curious if you have collaborated with or know of any research that scientists have done on projects you've designed to see how they're performing in terms of their ecological functions?

PB For my vertical gardens you can imagine that the research is mostly about insects, spiders, and birds. But I think the most interesting would be to know about the substrate, which is a very thin layer comparable to the mosses covering the rocks and forest. It would be very interesting to have a thorough study of all the microorganisms—bacteria, fungi, the unicellular algae, the mosses—all of this life in a three-millimeter-thick substrate. I really hope that someone will take on this research to learn which microorganisms arrive spontaneously on these vertical gardens. There is still a lot of work to do—not just in terms of the animals that the plant life supports—but in terms of what arrives in the vertical substratum to support the life of the plants.

CM I think in our case it's much more linked to the power of the place to make a more sensitive link between the people and their environment. People are afraid to be in contact with wildness and with life—with the unexpected—but if we don't have that connection, we lose. The landscape architect, the botanist, and the researcher have the power—actually it's a responsibility more than a power—to create places that are open to behaviors and practices other than what we are familiar with. It's the only way to evolve. We are lucky that we have many cultural briefs where we have a lot of freedom to direct the project. We are authorized to do things that are not done in usual cases with strong regulatory frameworks, like in Paris or New York or other big cities, which result in normative projects. It's a bit sad because you don't learn anything anymore. We do not work for our generation, we work for the next generation, but it takes a long time. One life is not enough.

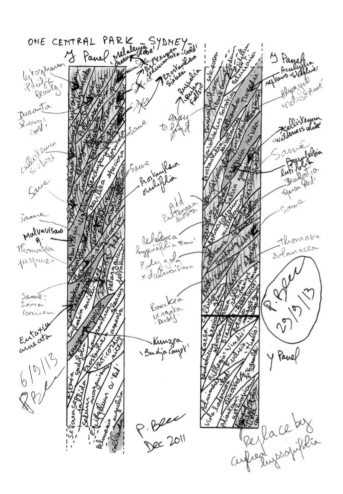

Below: One Central Park, Sydney, Australia.

+ Catherine you mentioned future generations and Patrick you spoke about what the plants need to survive. Have your plant choices or thinking about plants changed given what is known about the pace and effects of climate change?

PB No. First, as I always say, the most important thing is to stop destroying the natural world because once you destroy the habitat, you destroy the species and then it has no chance to adapt. To adapt to the climate is a luxury, it means you are still alive. The only real battle we need to have is to protect the remnants that are left. I recently visited the Western Ghats forests in India where maybe 5–8% of the forest remains. Most of these species will disappear before they get a chance to adapt. So, in my opinion, it is not important to know how they will change their habit, we just need to give them the space to change and to keep alive what is alive today.

CM We cannot change the world just in one country, or one project, but each small thing can point to a way to change direction. We know very precisely the scientific news and what we need to do, and we have to do it together. We, as landscape architects, can play a small part in changing direction by showing what is possible. I am an optimistic person. It's much more powerful than just repeating what has been done before. But that means we should also be aware of research in a broad sense, not only contemporary works but also looking to history—I recently learned a lot about the power of plants' effects on the water cycle beyond just the liquid phase—for how scientists of the past thought through issues that are relevant to us today.[2]

+ What information should be required knowledge for emerging landscape architects? The knowledge base is so vast—it is something that is learned gradually—but how do you approach it within your office in terms of educating young landscape architects into thinking about plants?

CM That's a difficult question because my primary background before I was in landscape is biology, physics, and chemistry. I am always interested in such level of knowledge. Then, as I mentioned, we made *Pages Paysages* for 10 years. So I was lucky to cross different kinds of knowledge and different kinds of experimentation, which is really important because whatever we do as designers is always an interpretation. It is not enough just to know plants. Everything we do is cultural. I lose my patience when the focus is only on ecology because even ecology is a cultural vision. So, we

+ I don't know the situation in France right now but in the US there seems to be a waning interest in botany as a field of scientific study. The US government is encouraging people to apply to enter the field.

PB In France, the study of botany is, unfortunately, declining and there are very few courses. Most of the botanists are between 40 and 70 or 80 years of age. But globally, the story is very positive. There is a new generation of botanists, most from tropical countries, and there has been an increase in publications written by local botanists. Why? Because during COVID, the European and American botanists did not go in the field, so the local botanists worked by themselves for three years. And now there is a burst of publications, and this is great. Botany, especially tropical botany, has come back to the countries where plants have been studied for 200 years by botanists from other countries.

+ One final question, are there any particular projects that you are working on that you're excited about?

CM I am in the process of codirecting a large volume about landscape architecture in France focused on the last 30 years of practice. We think France has a very singular approach, and a large number of practices, but we feel we are not so visible. It's ambitious because we are so many people. There are approximately 70 authors – it's a manifesto, really. It will be three volumes and, like *Page Paysages* and *LA+*, it's not only landscape architects, but also philosophers, sociologists, historians, and others.[3]

PB Yes, many. I do projects of all sizes–from interior private gardens, to airports, malls, and museums–and in many places. I continue to learn many things. I recently returned from India for a project at the Bangalore airport and was so surprised by many plant species. I didn't know the counterpart species in Southeast Asia. The same genus has a totally different growth habit and adaptations to different types of supports. What is important, always, is to have the opportunity to travel. Everywhere is the chance to be a botanist. When you are a zoologist, it's much more complicated because animals move. Plants don't move so it's easy to make observations. You have just to open your eyes.

---

1 Patrick Blanc, "Etre Plante," *Pages Paysages Anamorphose* 7 (1998–1999): 96–101.

2 For example, John Woodward's "Some Thoughts and Experiments Concerning Vegetation" (1699) and Stephen Hales's "Vegetable Staticks" (1727) are two famous botanists who studied transpiration and forester Viktor Schauberger (1885–1958) who developed unique theories about the movement of water.

3 Landscape Collectif. The editorial board is Caroline Bigot, Marc Claramunt, Philippe Clergeau, Jacques Coulon, Denis Delbaere, Yves Lugimbühl, Catherine Mosbach, and Sylvie Salles. The first volume of the book– titled *Le projet de Paysage, un Manifeste Critique* (Hermann)–will be published in 2024.

Opposite: Taichung Central Park.

CONCEITS AND CONSTRUCTS:
VEGETAL ARCHITECTURE

ANNETTE FIERRO

Annette Fierro is an associate professor of architecture at the Weitzman School of Design at the University of Pennsylvania, where she was associate chair of the department between 2016 and 2022. Her teaching and research address issues of technology within contemporary international architecture and urban culture. Fierro is author of *The Glass State: The Technology of the Spectacle, Paris 1981-98* (2006) and her most recent book, *The Architecture of the Technopolis: Archigram and the British High-Tech*, was published by Lund Humphries Press in 2023.

+ ARCHITECTURE, TECHNOLOGY

The discourse of architecture has evolved through one very long and essentialist polemic: building positioned against nature – as a separation from it, as a conquest of it, as a reconciliation with it, and, in its latest iteration, as a tightly bound symbiosis with it. These relationships have long been literal as well as metaphorical, technical as well as philosophical. The integral dialectic between architecture and nature originates with the construction of the first shelter and spans the millennia to an array of contemporary experimental dalliances with science and engineering. Within this spectrum there are degrees in which the status of the architectural object has been positioned to approach and even dissolve into either botanical phenomena of nature or its organizational and/or dynamical attributes. In this exchange, the role of "image" is decisive. For as much as nature itself has been continuously refigured by human intervention within the sphere of architecture and landscape architecture, as theoretical construct and reality, nature has also been continuously refigured through imagery.

Currently, in architecture schools and practices, gauzy veils of botanical matter have become generic. Software rendering packages, from Photoshop to ZBrush, insinuate that any building might be conditioned by soft pastoral foregrounds, despite the actual climatic and environmental conditions of the site. Texture mapping techniques suggest that surfaces are infinitely malleable and open to refiguration by material growth and decay. Digital scripting and artificial intelligence indicate that evolutionary models or even randomly posed phrases might simulate wildly reconstructed permeations and visionary objects. Contemporary architects David Ruy and Karel Klein have noted that their early work "out-natured nature."[1] As many have acknowledged, nature as wild and untouched is as fictional as the more fantastic of the images of it, yet the desire to locate, symbolize, simulate, and ultimately *recover* nature is one of the most fundamental of unrequited goals of architecture and the many forms of imagery it entertains.

The urgency of the climate crisis has spawned its own set of architectural responses. A discussion of real technologies coping with climate change is out of the scope of this discussion. Rather, this essay considers the notion of the finite, tangible building populated by planted matter–"vegetal buildings"– and it considers the agency of the image of these buildings as equivalent to their ecological function. Throughout this essay, the term "image" is used broadly to refer to the different ways that the relationship between nature and architecture is configured, from drawings, to built works, to photographs of built works. For at least two decades, architect Jean Nouvel has actively challenged the architectural object as codependent on literal nature in dozens of built and unbuilt projects at considerable scale, broaching urban dimensions. For Nouvel,

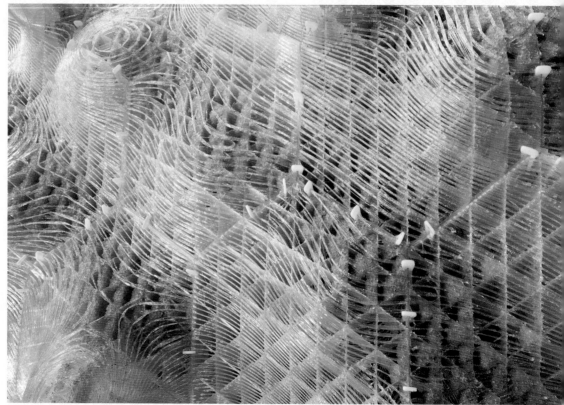

an image is a provocation, not just a view of how a finished building or landscape is presumed to look. His and others' contributions to President Sarkozy's 2009 competition *Le Grand Paris* popularized visions of vegetal urbanism into mainstream Parisian vernacular, subsequently supported by the urban policy of an environmentally motivated mayor. Global architecture in the same vein has emerged simultaneously. SOM's recent master plan for the central area of the Guangming District in Shenzhen features parametrically designed skyscrapers replete with forested terraces dripping foliage, surrounded by agricultural fields and vast parks. "Vegetal Cities" has become a term that has inspired several books and journals and is now entering the domain of urban policy. Once seen as eccentric in the 1970s and '80s in works by Peter Cook, Jean Renaudie, Ot Hoffman, Emilio Ambasz, and Friedensreich Hundertwasser, recent projects have vastly expanded the repertoire of vegetal building beyond idiosyncrasy. As spectacles, the credibility of visions that incorporate vegetal matter as building elements rely on suspended disbelief – their appeal is fanstastic, relying on the impulse to regard them as utopic, and yet the fact that several buildings have been built suggests that they are indeed possible in political, economical, and hydrological terms. Following Frederic Jameson, it is my contention that the power, and thus agency, of the imagery is in its suggestion of an unrequited dream.[2] This is true even in cases of considerable scale that have incrementally tested new technologies, from Jean Nouvel and Edouard François's innovations in France, to Kengo Kuma's buildings in Japan and China, to Stefan Boeri's Bosco Verticale in Milan, to–invading usually prosaic American contexts–MAD Architect's housing tower in Denver.

This brief essay interrogates the underlying content of these emergences, and establishes a rubric of issues to understand their power, especially as images. First is a discussion of the vegetal surface as one born into literality from the long history of ornamentation, which sheds its signifying role and becomes real, living matter; second is the notion of the building's surface as an environmental interface, a membrane dynamically born of technological factors, and one tied historically to the extermination of ornamentation; and third is the image of building overcome by a greened landscape as a political figure of social equity. These three facets also call forth multi-scalar

Above: Ruy-Klein, David Ruy and Karel Klein, KLEX, 2008.

Below: Eco-Logic, Claudia Pasquero and Marco Poletti, H.O.R.T.U.S., Mori Museum Exhibit, Tokyo, 2022.

technological enterprises, from the artisanal crafting of surfaces to urban environmental infrastructure, and from building envelopes that unite environmental systems–traditionally thought of as impermeable–to far greater ecological agencies. Though a full discussion of each topic is outside the scope of this essay, this "setting forth" should be regarded as an introductory explication of this curious phenomenon of fuzzy vegetal buildings.

Ornament and Its Artisans

Edouard François has said that vegetal building, of which he is a prolific author, was born of modern architecture's eradication of ornamentation. Indeed, the turn to active planting on a building's surface is, he says, "obvious."[3] Rather than the crude naïveté depicted by Adolf Loos at the advent of modernity, the ornamented surface is a deeply historical phenomena rooted in perhaps most simply, artistic drive – the sensory pleasure of elaborating surfaces with patterns and sculptural forms, evolving into systems of narrative, symbolism, and even semiotics over millennia. Ornament is also deeply tied to representations of nature. Ornament, meaning "something that lends grace or beauty"–and the "act of adorning" itself–is a manipulation of a natural substance in the interest of putting our human stamp upon it.[4] It is no surprise that ornament is often botanical, which can be traced in lineages of meaning. The palmetto–a symmetrical set of fronds emanating from a single stem–in ancient Egyptian architecture evolves into the acanthus in the Greek and Roman Corinthian column and resurfaces in the graphic overlay of Herzog and de Meuron's Ricola Storage Building (1987).[5] Originally a symbol for the sun and harmony evolves over 3000 years into Swiss herbology. The *fleur-de-lis* has a similarly transcendent evolution into more extreme forms of authority. For William Morris and the Pre-Raphaelites, the return to floral ornamentation had the connotation of a return to artisanal labor as a protest against industrialization.

Addressing contemporary digital architecture, Antoine Picon links traditional ornamentation to human subjectivity and politics, and notes the allergy to any similar role of meaning in its digital reemergence within the discipline today.[6] He also notes that the digital project changed the notion of applied ornament into immersive surface, a notion close to arabesque decoration in Islamic architecture, which united geometry with the botanical in the multi-layered incrustation of entire buildings. This degree of ornamentation is highly dependent on craft, from master stone carving, to cast-iron pouring, to more recent digital 3-D printing, all of which afford exacting detail. Comparable qualities are the finesse of the texture, the intrinsic motion implied by the imagistic forms of nature "growing" across the otherwise bare flank of the building. The challenge to the status of the inert object is evident as an incorporation of a larger metaphor – just as leaves enable motion and growth, so too are they signifiers of them and indicative of larger ontological shifts.[7]

The latest of the experimental work in architecture has made its way from small investigations in "boutique" offices into sizable building elements from a variety of large-sized firms. These typically appear as monolithic rain screens–elements without the capacity to function technologically as a separation between interior and exterior environments–seen in such examples as Diller Scofio Renfro's The Broad in Los Angeles (2015) and Snohëtta's Le Monde Headquarters in Paris (2022). These explorations are sculptural, and the centrality of their conjecture is aesthetic. These enclosure systems, developed since the mid-1990s, rarely incorporate orientation, much less site context. However, there is experimental work where nature remains a deep field to be plumbed, and in which there is incorporation of and feedback from actual biological matter and material. These are born of ornamentation's aesthetic dimension–beautifully wrought surfaces–but are endowed with finely tuned material responses, where living matter makes the leap from mere representation to actual biology. Examples include Achim Mendes's exploration of climactic affect on materials, Claudia Pasquero and Marco Poletto's (EcoLogic Studio) incorporation of living matter into complex geometric structures, and Dana Cupkova's (EPIPHYTE Lab) design of building components that encourage biodiversity. Also within this experimental sphere are Neri Oxman and Mitchell Joachim's (Terreform One) active cultivation of biological growth, at all scales. These are all genuinely scientific in their experimental integrity. As much as they promise enormous development for a deep symbiosis of building and organic matter, they also invoke interesting shifts between representation, ornamentation, and image, working through all three simultaneously.

Interface and the Crisis of Image

In terms of typologies, the buildings most closely associated with botanical matter are, of course, greenhouses. In contrast to the discussion above on ornamentation, there is little to no simulation of botanical imagery in these buildings, which followed the advent of cast-iron and glass construction in other industrial forms in casting off everything but technological systems. The precise registration of simulated climate is particular to greenhouses. Answering the needs of environmentally estranged botanical specimens, these plain glass and steel boxes are outfitted with intense environmental systems and do not attempt to express anything other than a pure form of technological manifestation.

It was from one clever detail in the Crystal Palace (1854), itself a type of greenhouse, that Joseph Paxton realized the benefits of modularized construction. Cast-iron departed from its artisanal ranks and initiated all kinds of modern industrial processes: mechanization, mass production, prefabrication, standardization, systems-integration, and rapid site assembly, all in a simply enclosed glass envelope. Undoubtedly the greenhouse is one of the basic prototypes for glass and steel construction – it is a proto-tenet of modern architecture, and correctly mute semantically. One hundred and thirty years later, the technology of the greenhouse was also responsible for the development of structural glass. In the enclosure for the Grands Serres (Cité de Sciences, Paris, 1986), engineer/architects Rice Francis Ritchie (RFR) realized that the dynamic forces latent in the surface of tempered glass could be mobilized.[8] RFR's potent design was all things dynamic, the glass and its support structures presented a *mapping* of natural forces: from structural glass panes hanging from each other, to a network of lateral braces and universal joints that allowed the entire system to move horizontally.[9] And yet the appearance of the glass box remains inert, inscrutable to most, sequestered safely within its modernistic trajectory. The question of image as it pertains to the greenhouse cannot, however, discount the presence of the plants within. If the envelope of glass is the ultimate barrier suspending the growth of the plants, it forms an environmental, but also a visual, counterpart to them. It is a jewel box – its dependents pressed against its boundaries, struggling to emerge into an environment in which they could never survive. The inner volume of an artificial exotic climate is perceived as a two-dimensional compressed surface, a vitrine showcasing the foliage within. Against a precise statement of the imperatives of pure technology, it is the composite image—of the building with the plants—that endures. This too anticipates the building's surface as one composed of actual vegetal matter.

If we turn to Herzog and de Meuron we see in their audacious application of photographic images onto the exterior surfaces of their early buildings a canny statement recapitulating the dilemma seen in the greenhouse as being between image and non-image. By applying highly graphic two-dimensional imagery of a stylized palmetto stem as a film layer onto the polycarbonate enclosure of their Ricola Storage Building, they reinvent the modern glass box. As one of many architects in the post-modern confrontation against the impoverishments of modern architecture, these have much in common with Venturi Scott Brown's early iconic architecture. In a similarly one-dimensional statement, this modest building brims with centuries of architectural irony, repositioning image and ornamentation as architecture's fundamental components. This particular building, emphatically underscoring the potency of botanical imagery, also touches upon deeply rooted current sympathies. In 1984, Edward O. Wilson proposed a theory that might have seemed self-evident – human beings have an innate (biological) affinity for natural life forms.[10] Countless studies in neuroaesthetics in the last few years have indeed proven that our psychological and physiological well-being is affected strongly by contact with nature. Most importantly, this pertains as well to "artificial nature," from artificial plants, to wallpaper, to images on billboards and buildings in the urban surroundings. Image on its own, it seems, offers verifiable benefit. And it emphatically deserves our attention.

Radical Ecologies of Vegetal Building

With the themes foregrounded above, I turn finally to the vegetal buildings introduced earlier. My contention is that these buildings, often disdained as mere images of "greenwashing," operate across a wide range of issues. This is in addition to that which is at the assumption of their popularity – a response to the urgent climate crisis. A few questions in this ongoing research remain.

Is the notion of image critically addressed by the architects involved? There is a great variety in these architects' works. In Jean Nouvel's œuvre, the answer is resoundly yes. Beginning with the Fondation Cartier and continuing with Musée de Quai Branly, both in Paris and done in conjunction with the painterly botany of Patrick Blanc, Nouvel has deployed imitations and simulations of nature strategically to endow his urban projects with challenging relationships to nature, in both aesthetic and cultural terms. At Branly, the central garden is a simulacrum of colonialist landscapes, mirroring the collections inside. In his work in the early 2000s, this literality is hyperbolized in two variations. In the first, wholly simulated hillscapes in his projects are exemplified by the Museum of Human Evolution in Burgos, Spain and the Tokyo Guggenheim Museum. The radical degree of artificiality of nature is itself the primary characteristic of the work – these are deliberate provocations. In the second, we see more variants that he calls "landscape objects," which are replications of a natural form rescaled into their own natural contexts – a "desert rose" for the National Museum of Qatar, a mimicked Grand Canyon for a Las Vegas hotel project, and an icy rock "landscape object" in Iceland. Nouvel has also played consciously with the images of phantasmagorias of nature in interior spaces, proposing sensationalized gardens, manifested in the spectacularly intricate sky/ceiling of the Louvre Abu Dhabi. Finally, in Nouvel's return to Paris–in his competition proposals for the renovation of a central Parisian park at Les Halles and his entry to *Le Grand Paris*–projects are framed around the notion of a fecund garden as civic public space in response to the crisp geometric formality of a traditional French garden.

Do vegetal buildings function environmentally and ecologically? The metrics of environmental and health impacts are still highly speculative. These pilot projects are enormously complex and dependent on scales eventually yet to be realized. Antoine Picon has noted that any of the large-scale visions will be highly dependent on dwindling resources, of water especially.

Uncontestable, however, is the decrease in temperature within both urban and individual domains that forested surface will prompt, and an increase of biodiversity within cities, especially of birds and their accompanying ecosystems.

How have images of greened buildings been mobilized as an urban project in multiple dimensions? The notion of a greened utopia has loomed large in many urban arenas since Ebenezer Howard's Garden City, where the image of a green landscape was accompanied by an operational social vision, and this pairing has long been one in the public imagination. If nature imported at a city scale is first and foremost a social endeavor, its signification in visionary efforts is immediately legible. *Le Grand Paris* comprised proposals for environmentally proactive measures but was also an effort to bring the city's peripheral communities into the whole of Paris proper, embracing long-problematic questions of identity and shared resources. The greened images became emblematic. Although the original competition entries had little possibility of ever being realized, the second round of the competition for the new territories of the city were redundant with images of fecund nature overwhelming the built environment, now with realizable potential. It is also quite possible that the celebrity of *Le Grand Paris* played a considerable role in inspiring the initiatives of the current mayor, who has called for vast new treed urbanscapes and greened rooftops across the city. As part of this, Hidalgo's "Reinventer Paris" of 2014 proposed reinventing 23 public spaces and buildings with aggressive ecological policy, one of which was David Chipperfield's Morland Mixité Capitale (2022) on the right bank of the Seine. This project combines low-income housing with a five-star hotel, a rooftop community garden, a market, and bicycle rental. Kengo Kuma, Sou Fujimoto, Jacques Ferrier, and Tryptique Architects with Philippe Starck have all proposed vegetal buildings in Paris, some within this initiative.

Several architectural projects by Edouard François, as well as Italian architect Stefano Boeri, demonstrate that towers–symbols of urbanism–can paradoxically prove to be provocateurs of nature, pointing specifically to the agencies of images. Encounters with Boeri's Bosco Verticale have been described as a shock, the vegetal tower emerges incongruously from among banal speculative housing towers.

Above Left: Rice Francis Ritchie with architect Patrick Berger, Glass detail, Parc André Citroën, Paris, 1992.

Above: Charles Rohault de Fleury, La Serre de Nouvelle Calédonie, Jardin des Plantes, Paris, 1834-6, last of a progression of greenhouses since 1717.

Below Left: Herzog and de Meuron, Ricola-Europe SA Production and Storage Building, Mulhouse- Brunstatt, France, 1993.

Architect Andrew Todd says: "Boeri's Bosco Verticale towers are scrumptiously Instagrammable, and I was thus primed to dislike. At first sight–in the arid Porto Garibaldi context–I immediately revised my view: the eye roams and consumes by chunks of character."[11] The shock seems to have worked – Boeri has been commissioned to work on a masterplan for Rome. Boeri also claims that the variety of plantings in his Bosco has spawned the nesting of 20 species of birds.[12] François says that his buildings first and foremost call into question the local and extended context of the building. François's residential Tower Flower–with its giant terracotta flowerpots on its balconies–plays with images of domesticated nature, while providing its inhabitants with natural light filtering through foliage. Planted with native species, François's Tour de la Biodiversité in Paris becomes a seedbank, allowing the wind to disperse seeds to the tower's immediate environment, regenerating the regional natural landscape. François has also developed his own prototype for housing in which that symbol of social hierarchy–height–is inverted. Each inhabitant is provided with a sizable garden terrace, the (subjugated) lower floor inhabitants are granted the highest of the gardens, the privileged upper floor inhabitants are given the lowest of the gardens, thus using the placement of the garden terraces as a mixing ground for social equity.

Plants, for François, recuperate the full dimension of context. "Context" includes the local and expanded constituency of site, but also spans several branches of architectural history, from ornamentation, to representation, to urban utopias and their many social and political concerns, as well as actual ecological, physiological, and psychological functions in both human and nonhuman spheres. The dual agencies of imagery and ecology coexist in vegetal buildings. As more buildings and urban proposals are attempted and realized, the possibilities and questions posed by these botanic visions will certainly expand.

---

1 From Ruy Klein's lecture at the Architectural League's *Emerging Voices* series (March 23, 2012).

2 Frederic Jameson, "The Politics of Utopia," *New Left Review* 25, no. 25 (2004). Jameson tells us that one defining character of any utopia is that it is destined to be unrequited, but that the draw toward that unattainable dimension is so strong that he calls it libidinous.

3 Interview with author (November 22, 2022).

4 Michael Mosteller, "Louis Sullivan's Ornament," *ArtForum* 16, no. 2 (October 1977): 44–49.

5 Alois Riegl, *Stilfrage: Grundlegungen zu einer Geschichte der Ornamentik* (1893), trans. by Evelyn Kain as *Questions of Style: Foundations for a History of Ornament* (Princeton Legacy Library, 1992).

6 Antoine Picon, *Ornament: The Politics of Architecture and Subjectivity* (Wiley, 2013).

7 Dario Gamboni, "Art Nouveau, The Shape of Life," in Angeli Sachs, ed. *From Inspiration to Innovation: NatureDESIGN* (Lars Müller, 2007).

8 The Grand Serres is named after the Serres of the Jardin des Plantes at the Cité des Science et L'Industrie in Paris.

9 For an in-depth explanation to this constructional system, see Peter Rice & Hugh Dutton, *Structural Glass* (Taylor & Francis, 1993).

10 Edward O. Wilson, *Biophilia: The Human Bond with Other Species* (Harvard University Press, 1984).

11 Interview with author (December 2022). Andrew Todd is an architect and essayist in Paris.

12 Stefano Boeri, "Un Manifesto: Imparare dal primo Bosco Verticale," from "Urban Forestry," https://www.stefanoboeriarchitetti.net (accessed December 18, 2022).

Opposite Above: Jean Nouvel (AJN), Jean-Marie Duthilleul (AREP), Michel Cantal-Dupart (ACD). *Le Grand Paris*, 2009.

Opposite Below: Atelier Jean Nouvel with Patrick Blanc, Vegetal and glass façades, Musée de Quai Branly.

SAMPLES

KAREN M'CLOSKEY

Karen M'Closkey is co-founder, with Keith VanDerSys, of PEG office of landscape + architecture and associate professor of landscape architecture at the University of Pennsylvania Weitzman School of Design where she co-directs the EMLab. M'Closkey and VanDerSys are authors of *Dynamic Patterns: Visualizing Landscapes in a Digital Age* (2017), guest editors of *LA+ GEO* (2020) and *LA+ SIMULATION* (2016), and editors of the forthcoming book *Media Matters in Landscape Architecture*. She is a fellow of the American Academy in Rome.

+ DESIGN, TECHNOLOGY, MEDIA STUDIES

A vast network of material and informational exchanges link landscape architecture and botany. As we scan nursery catalogs to select our favorite plants for a proposed design, much is forgotten about the seedy histories of colonization in which plant trade, bioprospecting, and botanic gardens played a central role. This network was dependent on a wide range of technologies that enabled the circulation of plants globally during Western colonial expansion – plant presses, Wardian cases, greenhouses, and others. However, given the unreliability of transporting living plant material, and the lack of detail and color in pressed specimens, botanic illustrations were an essential component in the transmission of botanic knowledge. While the circulation of images may seem inconsequential compared to the movement of plant material, historians have shown otherwise. Images were as much a part of the rise of botany as were the collection of plants and development of standards for plant identification and naming. As Daniela Bleichmar states, "In their function as mediators between the field and the cabinet, [botanic] images not only embodied and transported observations but also carried out multiple erasures – of place, of distance, of time, of human actors."[1] This essay considers how recent technologies have put new kinds of botanical images in circulation, namely the "plant samples" housed in digital imagery databanks. Though these images are virtual, they are, as with earlier botanic images, no less material in their effects.

Three kinds of digital samples that shape the representation of plant matter are outlined below. These sample types are housed in either vast image databanks, in spectral libraries, or in object libraries. The first are used as training data for the development of artificial intelligence. The second are used to identify materials in remotely sensed imagery by detecting a material's spectral signature, which is its emittance or reflectance value measured in wavelengths. And the third grouping refers to 3D-modeled plants available within rendering software. While these three plant sample categories are distinct in terms of the models upon which they are based, and the uses to which they are put, they are, to borrow Bleichmar's phrase, "visual avatars" that render nature a transportable object divorced from locality and contingency.[2] This distancing is an unavoidable condition for any landscape image, digital or other; however, the speed and ease with which we can obtain and propagate digital plant samples belies the significant amount of interpretation that happens between the field (i.e., the source of the plant imagery) and its representation in a landscape view.

Deep Learning

Landscape architects probably do not spend much time thinking about artificial intelligence (AI), at least not the kind where robots take over the world, and our jobs. But AI operates in much less obvious ways than a Roomba vacuum, and its implications are far more significant than whether Netflix recommends movies I might like. Over the past few years there have been only a handful of articles written about the relevance of AI to landscape architecture, yet AI's impact in the art and graphic design world has been rapid with the development of deep learning algorithms.³ Deep learning is a recent and highly sophisticated subset of machine learning that mimics the structure of the human brain in terms of its associative abilities.⁴ These "artificial neural networks" have been trained on millions of image-text pairs that are the basis of a number of platforms now available for public use. Just in the last five years AI platforms such as DALL-E, Disco Diffusion, and Midjourney have been making waves for their ability to spit out complex, unique, and compelling images in a matter of seconds, using the input of only a few words. The intricacy of the resultant images has grown significantly over a very short period. By the time this article is published, these tools will likely have been fully absorbed into image-making in landscape architecture.

Some artists and designers have been working with AI to test its possibilities and foster their own creative process. Auction houses wasted no time capitalizing on the novelty of it – Christie's sold its first AI painting for $432,500 in 2018. Others have been more critical about the effects that these images will have on the art and design world, raising concerns over job redundancy, creative process, and copyright, to name a few. Another primary concern—the one most relevant to this article—is how the technology works because of the inherent, and inherited, biases in the datasets upon which it depends.⁵ ImageNet, for example, is a database containing over 14 million images of "objects," sorted into over 21,000 categories.⁶ The images, initially scraped from the internet, are digitally tagged with words (i.e., classified) by tens of thousands of workers from Mechanical Turk, Amazon's crowd-sourced labor site.⁷ Perhaps not surprisingly, the most problematic labels are those under the object category "Persons" and critics have reported on myriad problems, such as identifying someone's gender or race from an image, assumptions about who constitutes the label "nurse," "lawyer," or "criminal," and so on, and the fact that most of the data is from North American and European sources.⁸

Whereas Dall-E and similar platforms use text prompts to create novel images from the millions of images it has "seen," the process can be reversed. Kate Crawford and Trevor Paglen brought to light the consequences of classifying humans in their compelling artwork *The ImageNet Roulette,* which allowed people to upload photos of themselves to see how they were labeled.⁹ This was fun and funny for some but very disturbing for others, proving Crawford and Paglen's point.¹⁰ Yet even their project was not without criticism given that some of the images it used for training were from a dataset where participants had not given consent for their portrait to be used for purposes other than the research for which the photographs were taken.¹¹ This does not detract from the power of the work; however, it is a potent reminder that the desire to unearth the problems of what is buried in datasets can overlook other aspects of those datasets, that is, the metadata or, in this case, the "terms and conditions" regarding use of the source material. Advances in AI make it increasingly difficult, impossible really, for "end-users" like landscape architects to have detailed knowledge about the provenance of the data and images in which we are awash. And though the use of software that samples from millions of image-text pairs to construct a landscape view may seem innocuous compared to sampling images of humans or plucking a live plant from the wild and transporting it around the world, this is not necessarily true when that view is a map.

Landscape architects have been using images assisted by machine learning for decades through the production of land cover maps, which are made by classifying pixels to sort them into predefined categories such as forest, wetland, and so on. Though the type of algorithms employed in deep learning neural networks have not been broadly used as a method of image interpretation for land cover classification, it is only a matter of time.¹² And though land cover classification long predates computers and remote sensing, the exponential increase in these technologies,

1 Daniela Bleichmar, "The Geography of Observation: Distance and Visibility in Eighteenth-Century Botanical Travel," in Lorraine Daston & Elizabeth Lunbeck (eds), *Histories of Scientific Observation* (University of Chicago Press, 2011) 392.

2 Ibid, 392.

3 Mimi Zeiger, "Live and Learn," *Landscape Architecture Magazine* (February 12, 2019); Zaš Brezar, "Using Artificial Intelligence in Your Design Process," *Landezine* (October 26, 2022); Bradley Cantrell, Zihao Zhang & Xun Liu, "Artificial Intelligence and Machine Learning in Landscape Architecture," in Imdat As & Prithwish Basu (eds), *The Routledge Companion to Artificial Intelligence in Architecture* (Routledge, 2021), 232–47.

4 Artem Oppermann, "Artificial Intelligence vs. Machine Learning vs. Deep Learning: What's the Difference?" *Built In* (May 2, 2022).

5 Sigal Samuel, "A new AI draws delightful and not-so-delightful images," *VOX* (April 14, 2022); Neel Dhanesha, "AI art looks way too European," *VOX* (October 19, 2022); Khari Johnson, "DALL-E 2 Creates Incredible Images—and Biased Ones You Don't See," *WIRED* (May 5, 2022).

6 See ImageNet, https://image-net.org/index (accessed January 15, 2023).

7 On image labeling, see Developedia "ImageNet," https://devopedia.org/imagenet (accessed December 10, 2022).

8 The developers of ImageNet acknowledge, "China and India are represented in only 1% and 2.1% of the images respectively," Ibid.

9 Julia Carrie Wong, "The viral selfie app ImageNet Roulette seemed fun – until it called me a racist slur," *The Guardian* (September 18, 2019).

10 Kate Crawford & Trevor Paglen, "Excavating AI: The Politics of Training Sets for Machine Learning," (September 19, 2019), https://excavating.ai/.

11 Michael J. Lyons, "Excavating 'Excavating AI': The Elephant in the Gallery," *Zenodo* (September 19, 2020), https://doi.org/10.5281/zenodo.4391458.

12 For limitations on machine learning methods for mapping land use and land cover change, see Junye Wang et al., "Machine learning in modelling land-use and land cover-change (LULCC): Current status, challenges and prospects," *Science of the Total Environment* 822 (2022), 153559.

13 Aarti Gupta, et al., "Making REDD+ Transparent: The Politics of Measuring, Reporting, and Verification Systems," in Aarti Gupta & Michael Mason (eds), *Transparency in Global Environmental Governance: Critical Perspectives* (MIT Press, 2014), 196.

14 A. M. Baldridge, et al., "The ASTER Spectral Library Version 2.0," *Remote Sensing of Environment* 113 (2009): 711–15.

15 Susan K. Meerdink, et al., "The ECOSTRESS Spectral Library Version 1.0," *Remote Sensing of Environment* 230 (2019), 111196.

16 R. F. Kokaly et al., *USGS Spectral Library Version 7* Data Series 1035 (US Department of the Interior, US Geological Survey, 2017), https://pubs.er.usgs.gov/publication/ds1035.

accompanied by the development of numerous and varied algorithms for image interpretation is, according to some scholars, hastening the "technicalization" of our environment.[13] The goal of these technologies has always been greater accuracy—a more "faithful" representation of the real world—however, the challenge of translating complex natural entities into words and pixels will not go away because questions of interpretation are never merely technical.

Spectral Libraries

The creation of land cover maps and any work related to tracking landscape change are dependent on "spectral signatures," which are gathered through camera sensors that can detect light waves beyond human vision. This information can be collected from in situ (field), airborne, or satellite sensing. Spectral signatures are used to interpret remotely sensed imagery and can be translated into land cover maps since each material (or material grouping, depending on resolution) has its own spectral signature. Whether using data collected from field samples or from satellite sensors, the process of using spectral signatures for image interpretation is the same, albeit field collection can contain more detailed information about the entities that compose a particular habitat (due to more samples and more materials being measured at higher resolution).

When field collecting plant material, a number of leaves may be gathered from a single plant or from multiple plants within the same geographic area. These are then arranged together in the lab within 48 hours of collection and their reflectivity and emittance values are measured and cataloged. The USGS began compiling samples for its spectral library in the 1980s. One such database, ASTER, has been distributed 4,000 times to over 90 countries.[14] It was updated to the ECOSTRESS spectral library, which currently contains over 3,400 spectra of human-made and natural materials.[15] ECOSTRESS was developed to study the temperature of plants to indicate if and where drought is occurring. This information can be used to forecast likely locations of wildfires or for use in crop monitoring. Though ASTER and ECOSTRESS sense the earth from the International Space Station (60 m resolution), the images are interpreted *through* the lab-measured spectra of individual leaf samples that were field collected. There are currently 1,230 vegetation spectral field samples in the USGS and NASA libraries. According to the USGS, the spectral library "forms a knowledge base for the characterization and mapping of materials and provides compositional standards of importance to a variety of research programs."[16] These samples were originally gathered to answer specific scientific questions and their locations are representative of the research of those who compiled the spectral library. Yet spectral libraries are also important for ecosystem monitoring "on a global scale and in a broad temporal context."[17] In other words, spectral samples are used for interpreting remotely sensed imagery in areas other than where the original samples were collected; thus, a spectral signature stands in for—is representative of—all individuals that belong to that spectral signature. This is akin, perhaps, to the late 19th-century invention of "type specimens" to standardize plant identification.

Upon identification of a new plant species, a single pressed plant becomes the representative of the species for all time.[18] These so-called "type specimens" or "holotypes" are stored in herbaria around the world. As Lorraine Daston writes, botanists were under no illusion that a single specimen could represent the variability of plants that belong to a species. In fact, many botanists resisted the idea of type specimens because it went against what was clearly observable in the field – the immense variety and diversity of individual plants. Nevertheless, as Daston states, the type specimen became "the face–the

desiccated, flattened face to be sure, but still the face–that is attached to the name of a species, and on the permanence of that relationship depends the transmission of botanical knowledge amassed for centuries."[19] Daston describes the compression of species into one representative individual, along with its labeling, publishing, and referencing, as an "art of transmission" that "makes a certain kind of science possible."[20]

Just as botanical nomenclature and type specimens depend on standardized practices and representative samples to ensure continuity in the transmission of knowledge across time and space, the development of spectral libraries is "a critical step in the standardization and automation of remote sensing interpretation and mapping."[21] Though not a type specimen in the strict sense, the individual entries in spectral libraries are similar in that a sample–in this case a spectral signature–represents a species or grouping of species. As a mediation between the field and the landscape image, spectral libraries *make a certain kind of science possible*. While much knowledge has been gained through these technologies, much is also obscured or lost.

It has long been recognized that land cover maps can be problematic because they homogenize many habitats, uses, and cultures under a single word such as "forest." A comparison among eight global maps derived from satellite imagery found huge discrepancies in how much of the earth is defined as "forested" depending on the definition of forest used. This is not simply a technical problem of instrument sensitivity or image resolution – this is because "forest" means different things to different people.[22]

There have been numerous examples of the misalignment in forest definitions between those who live with, and make a living with, forests on the ground and those involved in conservation who observe from above. This includes missing areas of forest due to image resolution but also because farmers and foresters, for example, do not see trees the same way, which means that "ground-truthing" does not solve the problem of data accuracy as it is often presumed to do. There are many examples of such disparities. One study showed the mismatch between the government's definition of "forest" and that of the farmers in Rajasthan, India.[23] The foresters registered successful reforestation efforts because they included mesquite in their tree cover count, which is an invasive species that the farmers considered degraded land. Another study points to the failed efforts of imposing Canadian forestry practices on Nicaraguan farmers who had converted their cropland into tree plantations in order to partake of payments for carbon credits.[24] With crop failures due to the importation of improper practices, the farmers received lower payments than "promised" because their land was sequestering less carbon. These are not isolated cases.

According to NASA "until [forest] definitions get hammered out, it will be difficult to make proper land-cover assessments or conservation agreements related to greenhouse gas emissions, climate change, and biodiversity."[25] But this presumes that these definitions can get "hammered out" by standardizing what a forest is–based on characteristics such crown cover, canopy height, and biomass–so that consistent measurement can be used to track change. Even if there were a shared definition of forest, which there is not, the methods for measuring that forest via remotely sensed imagery are problematic. Though the impacts of image resolution on land cover classification are well known, more recent work has shown that the same high-resolution imagery can lead to widely variable estimates of "crown cover"–one of the primary values by which forests are measured–depending on how the "reference area" for interpreting the image is defined. Reference area is unrelated to

PLANT SAMPLES
110

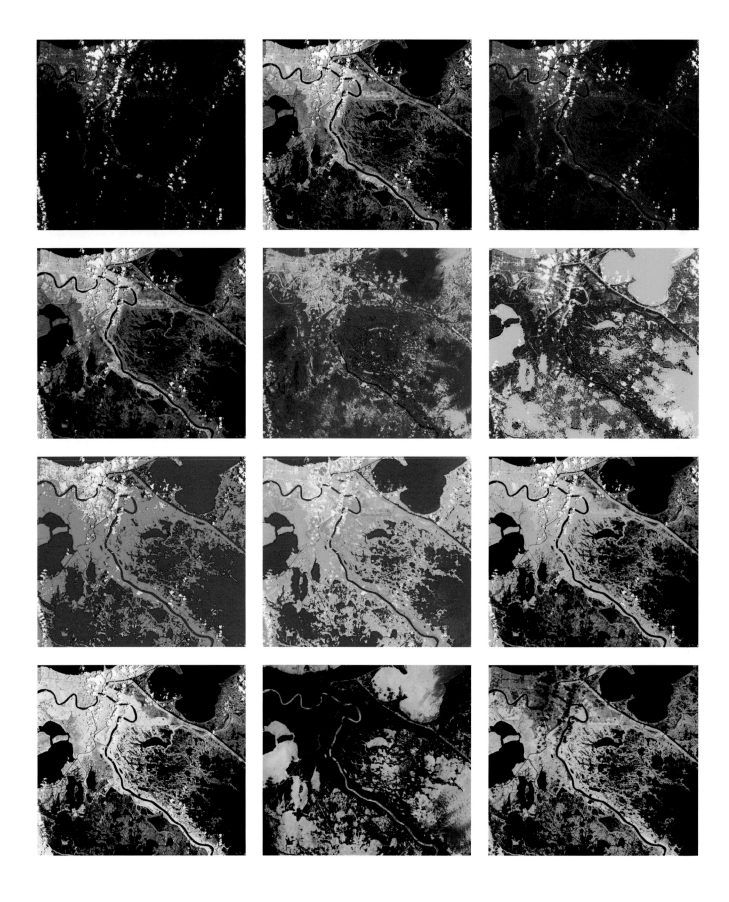

image resolution – it is simply the spatial extent over which the sampling of the image happens. For example, a 3x3 reference area is comprised of nine pixels and, as stated by those who produced the study: "the crown cover percent is determined in a reference area of a defined number of pixels around the pixel in question."[26] Their simulation for testing the effects of different reference areas showed that crown-cover estimates vary by as much as 50%–an astonishingly large number–which would significantly under- or overestimate how much forest exists.

Some remote sensing specialists suggest that algorithms that produce fuzzy sets, overlapping classifications, or use different mapping techniques that conceptualize forests as fields, rather than objects with strict boundaries, allow for a multitude of interpretations of "forestness" to be accommodated, and reflect the uncertainty inherent in mapping landcover.[27] Others suggest that the inherent vagueness is not a problem unless forests and other natural entities are linked to economic value, as is increasingly happening.[28]

Despite all of the known issues with divergent forest definitions, the climate crisis has led to a growing push toward precise measurement for use in carbon accounting and ecosystem services. The UN policy framework REDD+ (Reducing Emissions from Deforestation and Forest Degradation) is an important example. Satellite remote sensing and image interpretation using spectral signatures is what makes this program possible because it provides a way to measure forest change and thus is used to determine the payouts that countries receive for forest protection or replanting.

Though the + added to REDD program's second iteration was meant to value forests beyond their capacity to hold carbon– such as poverty alleviation and biodiversity conservation– it is merely exacerbating the problems inherent to many conservation efforts.[29] Opponents have referred to REDD+ as the last great land grab.[30] The program has negatively impacted Indigenous people because the forest definitions it uses can encourage plantation forestry, thereby producing more food insecurity and biodiversity loss. And while REDD+ allows countries to determine their definition of "forest" and is thus presumed to accommodate natural and cultural differences in forest definitions, state interests and local interests are not equivalent, as many of the program's critics have argued.[31] Landowners' rights are undermined as carbon buyers limit what can occur on the land, and Indigenous peoples who do not have title to the land are rarely included in forest management decisions. As Comber and Kuhn argue: "Mapped forests, however they are defined, are presented locally as 'facts,' which are then manipulated towards the interests of the state, excluding community views, perceptions and opinions from the classification and measurement activity."[32]

The virtual images made possible by spectral mapping may seem far removed from the technologies that enabled the circulation of plant material globally during colonial expansion,

[17] Leyre Compains Iso, Alfonso Fernández-Manso & Víctor Fernández-García, "Optimizing Spectral Libraries from Landsat Imagery for the Analysis of Habitat Richness Using MESMA," *Forests* 13, no. 11 (2022), https://doi.org/10.3390/f13111824.

[18] Type specimens did not occur until late 19th and early 20th centuries; however, botanists retroactively assigned type specimens to earlier botanists' collections by discerning their intent. Lorraine Daston, "Type Specimens and Scientific Memory," *Critical Inquiry* 31 (2004): 165.

[19] Ibid., 177.

[20] Ibid., 157, 155.

[21] Jingcheng Zhang, et al., "Machine Learning-Based Spectral Library for Crop Classification and Status Monitoring," *Agronomy* 9, no. 496 (2019): 1.

[22] NASA Earth Observatory, "Is That a Forest? That Depends on How You Define It," https://earthobservatory.nasa.gov/images/86986/is-that-a-forest-that-depends-on-how-you-define-it (accessed December 10, 2022).

[23] Paul Robbins & Tara Maddock, "Interrogating Land Cover Categories: Metaphor and Method in Remote Sensing," *Cartography and Geographic Information Science* 27, no 4 (2000): 295–309.

[24] This study is cited in Birgit Schneider & Lynda Walsh, "The Politics of Zoom: Problems with Downscaling Climate Visualizations," *GEO: Geography and Environment* (2019), https://doi.org/10.1002/geo2.70.

[25] NASA Earth Observatory, "Is That a Forest?"

[26] Paul Magdon & Christoph Kleinn, "Uncertainties of Forest Area Estimates Caused by the Minimum Crown Cover Criterion," *Environmental Monitoring and Assessment* 185 (June, 2013): 5,350.

[27] Robbins & Maddock, "Interrogating Land Cover Categories," 298; Alexis Comber & Werner Kuhn, "Fuzzy difference and data primitives: a transparent approach for supporting different definitions of forest in the context of REDD+," *Geographica Helvetica* 73 (2018): 152, 154.

[28] Magdon & Kleinn, "Uncertainties of Forest Area," 5,356.

[29] See the Joint Declaration of the Indigenous Peoples of the World to CBD [The Convention on Biological Diversity] https://int.nyt.com/data/documenttools/joint-declaration-of-the-indigenous-peoples-of-the-world-to-the-cbd-34/20b4fa27750039d7/full.pdf.

[30] Yolanda Ariadne Collins, "Colonial Residue: REDD+, Territorialization and the Racialized Subject in Guyana and Suriname," *Geoforum* 106 (2019): 33–47.

Previous: The left image shows forested landcover if forest is defined as an area with 10% tree cover as compared to the right image, which shows forest defined as 30% tree cover.

Opposite: Multispectral imagery of coastal region near New Orleans interpreted through different band combinations.

but it is important to recall what Bleichmar said about 18th-century botanic illustrations: *images not only transport observations but also perform multiple erasures*. The urgency of action that is required to change course on the crises of global heating and habitat loss can blind us to the problematic legacies of our environmental data and governance structures.

Plant Object Libraries

The final category of plant samples highlighted in this essay are those found in model-rendering software. A number of essays in the last decade have commented on the ubiquity of photoreal or "hyperreal" views produced by landscape architects to represent their proposed designs.[33] Such images have been critiqued for a number of reasons, including the use of verdant nature to camouflage the toxic legacy that lies within our landscapes, with little development in new aesthetic expressions for these "denatured" landscapes and little acknowledgment of the technical requirements needed for their transformation.[34] These types of views are often constructed with Photoshop but are increasingly made with model-rendering software like Lumion. There are many programs available – Lumion is highlighted for the simple reason that its use among students has become pronounced over the last few years and it has quickly been adopted by the profession.

Compared to the AI deep learning algorithms described earlier, which sample millions of images at once, the samples rendered in 3D models are limited to a small palette of "objects" contained in the rendering platform's library. Lumion's plant library contains approximately 550 objects. A quick survey of those that are available in "fine detail"– specimens likely to appear in the foreground of an image– shows a distribution across the globe in terms of plant origin, and the species are representative of various biomes. However, the total number of plants (550) and those available for fine detail rendering (46) is minuscule given there are 400,000 known plant species, approximately 35,000 of which are cultivated.[35] Furthermore, of all the specimens in Lumion's plant library, roughly one-quarter are labeled by genus only, 15 are classified as "generic plant," and 21 are classified as "weed," a meaningless designation unless understood within a specific context. When asked (via email) how they determined their plant choices, a Lumion representative replied that their selection is "commercially sensitive and we are not at liberty to discuss certain selection processes."

The ever-more lushly rendered and detailed images enabled by Lumion are composed of a ready-made plant palette. The apparent specificity of the plants within the images belies the fact that they are often stand-ins–visual avatars–for the actual species that would comprise a planting design but are not available from the software's limited plant library. Perhaps too much should not be read into Lumion's restricted samples; however, as with Dall-E and similar programs, the increasing hyperrealism of images is likely not matched by increasing knowledge about the complex entities and interactions that take place in the landscapes we seek to represent. Nor do we know the provenance of the plant samples that comprise our images. A recent article by Parker Sutton in *LA+ GREEN* notes how social media sites like Instagram may seem peripheral to landscape architecture but their "ubiquitous gaze is central to the way a generation of users perceives the landscape and structures their interactions with it."[36] In regard to the explosive interest in succulents trending on social media, which has a very real effect on the demand for these plants in the horticultural industry, Sutton notes that "Plants become autonomous objects that can be parceled out for our pleasure,

rather than members of a complex, interdependent web."³⁷ This is akin to Bleichmar's description of botanic images that "[define] nature as a series of transportable objects whose identity and importance was divorced from the environment where they grew or the culture of its inhabitants."³⁸

Conclusion

> The natural history illustration, with its flower always in bloom, its fruit permanently ripe, its animal suspended in clarity and permanence, was at once the instrument, the technique, and the result of natural history as a field of study.³⁹

The desire for increasingly accurate depictions of plant material was concomitant with the rise of modern botany. Scholars have shown how the circulation of images was fundamental to the diffusion of scientific knowledge while simultaneously concealing other forms of knowledge. The technologies behind the various plant samples described above also strive toward increasingly accurate depictions, and they too obscure what gets lost in translation through the classification of nature into discrete objects. While the automation of image making in AI platforms and the availability of 3D-modeled plants in rendering software may be innocuous compared to the spectral samples used for forest sensing and land cover interpretation, the various databanks and models that delimit much of contemporary practice are standardized and globalized. This is not to suggest that we avoid these tools altogether, but, as Bleichmar asserts, images both bridge and create distance. The digital plant samples described in this essay are useful surrogates for some purposes, but we should be mindful of how these tools are "disciplining" the practice of landscape architecture.

31 See for example, Gupta, et. al. "Making REDD+ Transparent"; Tracy Perkins & Aaron Soto-Karlin, "Situating Global Policies within Local Realities Climate Conflict from California to Latin America," in Julie Sze (ed.) *Sustainability: Approaches to Environmental Justice and Social Power* (New York University Press, 2018), 102–23.

32 Comber & Kuhn, "Fuzzy difference and data primitives," 152. Media scholars Jennifer Gabrys and Shannon Mattern have written on similar problems. See Jennifer Gabrys, "Smart Forests and Data Practices: From the Internet of Trees to Planetary Governance," *Big Data & Society* 7, no. 1 (2020): 1–10, and Shannon Mattern, "Tree Thinking," *PLACES* (September 2021).

33 See Karl Kullmann, "Hyper-realism and Loose-reality: The Limitations of Digital Realism and Alternative Principles in Landscape Design Visualization," *Journal of Landscape Architecture* 19 (2014): 20–31; Karen M'Closkey, "Structuring Relations: From Montage to Model in Composite Imaging," in Charles Waldheim & Andrea Hansen (eds), *Composite Landscapes: Photomontage and Landscape Architecture* (Hatje Cantz & Isabella Stewart Gardner Museum, 2015), 116–31.

34 Richard Weller, "The Innocent Image," *LA+ Interdisciplinary Journal of Landscape Architecture* 3 (2016): 14–15.

35 Korous Khoshbakht & Karl Hammer, "How Many Plant Species are Cultivated?" *Genetic Resources and Crop Evolution* 55 (2008): 925–28.

36 Parker Sutton, "Trending Green: Landscape in the Age of Digital Reproduction," *LA+ Interdisciplinary Journal of Landscape Architecture* 15 (2022): 27.

37 Ibid, 26.

38 Bleichmar, "The Geography of Observation," 392.

39 Ibid., 386.

SMART PLANTS
AND THE CHALLENGES OF MULTISPECIES NARRATIVE

URSULA K. HEISE

Ursula K. Heise is the Marcia H. Howard Term Chair in Literary Studies at the Department of English and Director of the Lab for Environmental Narrative Strategies (LENS) at the Institute of the Environment and Sustainability at UCLA. Her research and teaching focus on the environmental humanities, contemporary environmental literature and media, science fiction, and narrative theory. Her books include *Sense of Place and Sense of Planet: The Environmental Imagination of the Global* (2008) and *Imagining Extinction: The Cultural Meanings of Endangered Species* (2016), which won the 2017 book prize of the British Society for Literature and Science.

✚ LITERATURE, ENVIRONMENTAL HUMANITIES

Science fiction—as a literary genre that focuses sometimes on alien ecologies, at other times on futuristic environments, and yet others on a combination of both—has often featured plants in scenarios of space exploration or invasions of Earth by aliens. These stories usually follow recognizable narrative templates of the kind that narratologist H. Porter Abbott has called "masterplots," typical story patterns that are readily available in a particular culture and serve as the basis for cultural interpretation of real-life events as well as for artistic elaboration and variation.[1] Hostile take-over of human-owned territories or even bodies by plants is one such familiar narrative template, while benign hybridization of humans with plants is another – both often functioning as obvious metaphors of political antagonism, colonialism, or migration.

But a different, less familiar type of narrative stages human visits to alien planets where terrestrial distinctions between plant, animal, and human break down in novel ecologies that challenge the ethical principles that humans import from Earth. These challenges become most obvious in recent works of speculative fiction that take on board scientific discoveries about plant perception and communication as well as ethical arguments about the rights of plants, to the point of making plants three-dimensional fictional characters or even narrators. As these stories refigure plants from resources and infrastructures for the use of humans to intentional subjects and narrative agents of their own, they raise complex questions of multispecies ethics, justice, and narrative itself: Can plant-centered stories successfully strike the balance between preserving the otherness of plants and assimilating them into the human cultural artifact of narrative? What degree of intention and agency is it reasonable to attribute to plants? And what real-life implications do plant-centered narratives invite their readers to take away?

The masterplot of hostile plants—terrestrial or alien—taking over the world has a long history in speculative fiction. In John Wyndham's classic novel *The Day of the Triffids* (1951), for example—which has been turned into a feature film, three radio series, and two serial TV shows—carnivorous plants take advantage of a blindness pandemic to bring down human societies and take over the world.[2] In Frank Oz's satirical film *Little Shop of Horrors* (1986), Audrey II, the notorious "mean

green mother from outer space," thrives on human blood.[3] In M. Night Shyamalan's B-movie *The Happening* (2008) plants revolt against humans' ecological exploitation by exuding a toxin that induces mass suicide.[4] And in Olivia Vieweg's German graphic novel *Endzeit* (End Time, 2018; translated as *Ever After*, 2020), plants infect humans and turn them into zombies in a landscape of climate change and societal breakdown.[5] In works like these, the hostility of plants serves variously as a narrative premise for staging civilizational breakdown, invasion by ideological opponents, or the consequences of ecological degradation. As literary critic Heather Sullivan has shown, this tradition of portraying plants as vectors of infection, bodily invasion, and civilizational take-over can also be turned to account for progressive political visions: in Nnedi Okorafor and Tana Ford's comics series *Laguardia: A Very Modern Story of Immigration* (2018-19), the journey of Nigerian migrants—whose bodies contain human as well as plant genes—to the United States becomes a metaphor for a national, racial, and cultural hybridization that the story welcomes.[6] In both their paranoid and their celebratory versions, plant narratives such as these—beyond their obvious allegorical meanings—reflect on humans' ecological reliance on plants as well as on their purely instrumental use in modern, industrialized societies. As the science fiction novelist Ursula K. Le Guin once put it, "The relation of our species to plant life is one of total dependence and total exploitation."[7]

But Le Guin herself as well as other speculative writers and filmmakers have also used the genres of horror, fantasy, and science fiction to imagine different human-plant relationships and botanic futures. The comics series *Swamp Thing*, created by Len Wein and Bernie Wrightson in the 1970s, told the story of a scientist murdered by a rival and thrown into a swamp, where his body metamorphoses into a human-plant hybrid under the impact of fertilizer chemicals. This "Swamp Thing," a "muck-encrusted mockery of a man" combats villains even as he yearns to return to his human body.[8] But the remake of the original series by Alan Moore, Stephen Bissette, and John Totleben in the 1980s turns the story's implicit devaluation of Alec Holland's plant body upside down. In this version, a scientist called Jason Woodrue dissects Holland's body only to discover that it is not a human body metamorphosed into plant matter, but on the contrary plant matter organized into

---

1 H. Porter Abbott, *The Cambridge Introduction to Narrative*. 3rd ed. (Cambridge University Press, 2021), 52–56.

2 John Wyndham, *The Day of the Triffids* (Modern Library, 2003 [1951]).

3 *Little Shop of Horrors*, directed by Frank Oz (1986, Warner Brothers).

4 *The Happening*, directed by M. Knight Shyamalan (2008, 20th Century Fox).

5 Olivia Vieweg, *Endzeit* (Carlsen, 2018).

6 Nnedi Okorafor & Tana Ford, *LaGuardia: A Very Modern Story of Immigration* (Berger Books, 2019). The series *LaGuardia* was republished as a single graphic novel in 2019. For detailed analyses of Vieweg and Okorafor & Ford, see Heather I. Sullivan, "Cross-Infections of Vegetal-Human Bodies in Science Fiction," *Science Fiction Studies* 49, no. 2 (2022): 342–58.

7 Ursula K. Le Guin, *Buffalo Gals and Other Animal Presences* (Plume, 1987), 83. Italics in the original.

8 Quote is by Len Wein in Alan Moore, Stephen Bissette & John Totleben, *Saga of the Swamp Thing: Book One* (DC Comics, 2012 [1983-84]), 7.

9 Moore et al., *Saga of the Swamp Thing*, 100. Ellipses in the original.

10 Quote is from Moore et al., *Saga of the Swamp Thing*, 107-108.

11 Ursula K. Le Guin, "Vaster Than Empires and More Slow [1971]," in *Buffalo Gals and Other Animal Presences* (Plume, 1987), 115.

12 I have analyzed Le Guin's short story in far greater detail in Ursula K. Heise, *Sense of Place and Sense of Planet: The Environmental Imagination of the Global* (Oxford University Press, 2008), 17–21.

13 For a more detailed analysis of *Speaker for the Dead*, see Ursula K. Heise, *Imagining Extinction: The Cultural Meanings of Endangered Species* (University of Chicago Press, 2016), 228–31.

simulacra of human organs and limbs: human essence turns into contingent externality, while Holland's new plant identity moves to the center. "They wouldn't let me be human ... and I became ... a monster. ... But they wouldn't let me be a monster ... so I became a plant. And now ... you won't let me ... be a plant," Holland meditates.[9] That this discovery is made by the aptly named Woodrue or "Floronic Man," who defines himself as the "pain and bitterness of the woods! ... the regret and anger of the forests" at humans' environmental destruction, only reinforces this graphic novel's new interest in plant life and the plant's-eye view of the world.[10]

Le Guin takes the narrative portrayal of total plant-ness to a global level by portraying an alien version of Gaia in her short story "Vaster Than Empires and More Slow" (1971). As humanoid explorers visit World 4470, they find a planet inhabited only by plants, a "totally alien environment, for which the archetypical connotations of the word 'forest' provide an inevitable metaphor."[11] The grass-like and tree-like species fill them with increasing unease, both cognitively and emotionally, as they begin to explore them. They gradually discover that the entire plant sphere is globally connected via rhizomes, epiphytes, and drifts of pollen, giving rise to a networked planetary sentience that they are at pains to understand. The human presence fills this distributed intelligence with terror, given that it has never known anything or anyone except itself. Only the self-sacrifice of one expedition member, who chooses to merge physically and psychologically with the alien forest, allows them to continue their scientific investigation in harmony with their otherworldly surroundings. James Lovelock had proposed his influential Gaia Hypothesis of Earth as a super-organism, a planetary feedback system that sustains life itself, only a few years earlier. In "Vaster Than Empires," Le Guin offers a more unsettling version of planetary ecology as an incomprehensible plant intelligence that humans cannot dominate, and to whom they have to offer a voluntary sacrifice, in a reversal of the usual plant invasion plot.[12]

Sentient trees also play a central role in Orson Scott Card's novel *Speaker for the Dead* (1986), which is set on the planet of Lusitania – like World 4470 a planet whose ecology human settlers initially misinterpret. So-called "Pequeninos," the native intelligent and animal-like species, refer to individual trees by name. But only after the Pequeninos murder two of the human researchers, seemingly without motive, do the settlers discover the reason for this custom: the trees are not in fact a separate species, but the form into which male Pequeninos metamorphose after their death, an honor they meant to bestow on their human visitors. Distinctions between plants and animals as they pertain on Earth do not hold on Lusitania because of a virus that has scrambled genes for eons in the planet's evolutionary history, and that soon infects the colonists as well. The complicated ecological and medical plot that ensues constantly unsettles the humans'–including the readers'–sense of biological taxonomies and the ethics that are, at least in Western-influenced philosophy, assumed to go with them. What taxonomical status–intelligent being, animal, plant, virus–justifies putting an individual to death or exterminating a species, and what status protects it? Even more emphatically than Moore's or Le Guin's, Card's alien ecology highlights that biological classifications are never just that, but also cultural, moral, and legal categories that shape how a society understands justice.[13]

In spite of the freedom afforded by their choice of literary genres, these scenarios are not just the speculations of overly imaginative creative writers. Over the past decade, the Canadian scientist Suzanne Simard and the German forest manager Peter Wohlleben have presented evidence for systems of communication and interspecies collaboration among trees that scientists refer to as "mycorrhizal networks" and that popular language has dubbed the "wood-wide web." Following Wohlleben, philosophers Michael Marder and Paco Calvo have made the case for plant rights and plant sentience.[14] Some of these claims remain controversial for the time being: some plant biologists see no evidence of intention in the scent signals trees emit; some neuroscientists contest the possibility of intelligence on the part of organisms that do not have a central nervous system.[15] Practical ethical objections have been raised as well: some vegans see their arguments for animal welfare or rights and the ensuing dietary choices undermined by the claim that plants, too, are sentient life forms.[16]

But the fact that the status of plants is now being discussed in serious academic publications and public debates in ways that parallel similar conversations about animals after the publication of Peter Singer's *Animal Liberation* in 1975 shows that the emergence of plant sentience as a topic of science fiction is not just futuristic fancy but also a sign of shifting cultural perceptions of the plant world. Neither is this shift particular to the Global North: Indigenous cosmologies in many parts of the world have long emphasized kinship relations between humans and nonhumans, and in partial acknowledgment of this tradition, the new constitutions of Ecuador in 2008 and Bolivia in 2009 explicitly attribute legal rights to the "Madre Tierra" or Pachamama of Indigenous thought. Meeker and Szabari call this "radical botany" or, following Natasha Myers, "planthropology."[17]

The novelist Richard Powers has fictionalized Simard's and Wohlleben's wood-wide web in his Pulitzer Prize-winning novel *The Overstory* (2018), which narrates the cognitive ties between nine different individuals and particular tree species in the quite realist context of struggles against deforestation in the 1970s.[18] But it also includes italicized passages in which trees are portrayed as communicating with each other and with humans:

> Talk runs far afield tonight. The bends in the alders speak of long-ago disasters. Spikes of pale chinquapin flowers shake down their pollen; soon they will turn into spiny fruits. Poplars repeat the wind's gossip. Persimmons and walnuts set out their bribes and rowans their blood-red clusters. Ancient oaks wave prophecies of future weather. The several hundred kinds of hawthorn laugh at the single name they're forced to share. Laurels insist that even death is nothing to lose sleep over.
>
> *Trees even farther away join in*: All the ways you imagine us—bewitched mangroves up on stilts, a nutmeg's inverted spade, gnarled baja elephant trunks, the straight-up missile of a sal—are always amputations. Your kind never sees us whole. You miss the half of it, and more. There's always as much belowground as above.[19]

This imaginative scenario of active communication on the part of plants, which Powers develops on the basis of current science, still does not go as far as Sue Burke's science fiction novels *Semiosis* (2018) and *Interference* (2019).[20] Once again, humans land on an alien planet, Pax, where they soon discover that some of the native plant species are sentient, in another take on the scenario developed by Le Guin. They work with and put these plants to use, but an intelligent bamboo develops a system of written signs that allows him to communicate with humans. Yes, him: as the human protagonist points out, "[i]t earned a human name in honor of its importance, and pretty soon we started calling it 'him' as if he were a man and not a hermaphrodite plant."[21] The humans give him the name Stevland, which he accepts, and his structure of distributed intelligence, based on perception and cognition through his many individual shoots in different parts of the region, becomes a major source of support for the human settlement. He eventually becomes not just a citizen of the community, but co-moderator of its governing Committee – and steps down due to what he recognizes as his own misguided decisions after a time in office. Even as the humans continue to negotiate their doubts and ambiguities about co-existence, collaboration, and even co-governance with an intelligence so different from theirs amidst a multitude of alien species, Stevland undergoes his own development as he acquires fluency in human thought, language, morality, justice, and politics. Stevland the bamboo is one of the very few fully developed plant characters in fiction, and Burke goes further than any of her predecessors in combining a human perspective on plant life with a plant's perspective on human life – to the point where Stevland sends an ambassador offshoot back to Earth with human visitors, as perhaps the first plant astronaut in science fiction.

Le Guin extended her own experimentation with botanical narrative and plant character in her 1974 short story "The Direction of the Road," in which an oak features as the narrator. Standing by the side of a path that over time becomes a road and then an asphalted highway, the tree narrates technological progress from pedestrian locomotion to horse carts and motor cars as a feature of its own changing mobility skills: "I remember the first motor car I saw," it comments. "I approached it at a fair speed, about the rate of a canter, but in a new gait, suitable to the ungainly looks of the thing: an uncomfortable, bouncing, rolling, choking, jerking gait."[22] Its experience of modernity reaches a climax when a car crashes into its trunk, and the driver perishes with his dying gaze on the oak. The tree refuses the human gaze along with being cast as a symbol of eternity:

"I cannot uphold such an illusion. If the human creatures will not understand Relativity, very well; but they must understand Relatedness. ... I will not act Eternity for them."[23] In this rare instance of a multispecies narrative in which a plant itself takes over the narrative voice, it proudly rejects human meanings projected on it and demands instead the justice of ecological relatedness.

There is no doubt that it remains difficult to imagine how such narrative experiments in alternative relationships between plants and humans might shape material environments in the Anthropocene. For urban planners and landscape architects, plants remain an infrastructure to be created and maintained rather than an alternative society with intentions, agents, perspectives, and experiences of its own. For ecologists, plants primarily constitute habitat for other species. For farmers and florists, plants remain a commodity to be bought and sold. And for average citizens, they remain a basis of nutrition that it is—unlike animal food stuffs—impossible to forego completely. By way of a first step toward a new kind of plant ethics, a botanist character in Powers's *Overstory* proposes that "'when you cut down a tree, what you make from it should be at least as miraculous as what you cut down.'"[24] Native American botanist Robin Wall Kimmerer has elaborated the concept "reciprocity" in much greater detail as a way of turning humans' one-way relations to plants into a two-way gift economy – a reciprocity that is fictionalized in Le Guin's "Vaster Than Empires" as well as Burke's *Semiosis*.[25] Whichever theoretical concept or narrative scenario one takes as a launchpad, recent theoretical discussions about the sentience of plants as well as speculative botanical fictions invite us to imagine a community of justice that no longer excludes plants as subjects.

14 On plant rights see Michael Marder, *Grafts: Writings on Plants* (Univocal/University of Minnesota Press, 2016); on plant sentience see Paco Calvo & Natalie Lawrence, *Planta Sapiens: The New Science of Plant Intelligence* (W. W. Norton & Co., 2023).

15 Marder, *Grafts: Writings on Plants*, 480-547; Temple Grandin, "Can Plants Think? Review of Planta Sapiens: The New Science of Plant Intelligence, by Paco Calvo," *New York Times* (June 23, 2023), https://www.nytimes.com/2023/06/23/books/review/planta-sapiens-paco-calvo.html?searchResultPosition=1; Richard Grant, "Do Trees Talk to Each Other?" *Smithsonian Magazine* (March 2018), https://www.smithsonianmag.com/science-nature/the-whispering-trees-180968084/.

16 Marder, *Grafts: Writings on Plants*.

17 Natania Meeker & Antónia Szabari, *Radical Botany: Plants and Speculative Fiction* (Fordham University Press, 2019), 20.

18 Richard Powers, *The Overstory* (Norton, 2018).

19 Ibid., 3-4. Italics in the original.

20 Sue Burke, *Semiosis* (Tor, 2018); Sue Burke, *Interference* (Tor, 2019).

21 Burke, *Semiosis*, 142.

22 Ursula K. Le Guin, "The Direction of the Road [1974]," in *Buffalo Gals and Other Animal Presences* (Plume, 1987), 86.

23 Ibid., 91.

24 Powers, *The Overstory*, 452.

25 Robin Wall Kimmerer, *Braiding Sweetgrass: Indigenous Wisdom, Scientific Knowledge, and the Teachings of Plants* (Milkweed, 2013), 22-32.

Acknowledgments
I would like to thank my research assistant in 2022-23, Alik Shehadeh, whose outstanding work provided crucial background information and materials for this article.

IMAGE CREDITS

Endpapers
"The Night-Blowing Cereus" by Philip Reinagle (1807), public domain.

Editorial
p. 4: "An illustration of a Vegetable Lamb of Tartary" by H. Lee (1887), public domain via Wikimedia Commons.

The Changing Nature of Botanic Gardens
p. 6: Image composite by Maura McDaniel using "Kew Gardens Victoria Regia House" by unknown (1903), public domain, and "Image of Prince of Wales Conservatory" by Scott Wylie (2022) used under CC BY-4.0 via Wikimedia Commons.

p. 8: Image of plant tissue cultures by Lance Cheung (2013), public domain.

p. 9: "No fundo da Caldeira" by Jardim Botanico (2010) used under CC BY-SA 3.0 via Wikimedia Commons.

p. 11: Image by O. Luci (2013), CC BY-SA 3.0 via Wikimedia Commons.

Spiraling Diversity and Blank Spots in a 19th-Century Utopian Botanic Garden
p. 12: Drawing of Padova Botanic Garden, by Girolamo Porro (1591), public domain (altered), and drawing of "The botanic flower-garden with a gravel-walk" (1826) by J. C. Loudon used under CC BY-NC-SA 4.0 license (altered).

p. 14: Drawing from John Claudius Loudon, *Hints on the Formation of Gardens and Pleasure Grounds* (Gale, Curtis & Fenner, 1813) plate 17, public domain (altered on p. 12).

p. 16-17: Drawings from John Claudius Loudon, *An Encyclopedia of Plants* (Longman, Rees, Orme, Brown & Green, 1829), 836, 829, public domain (cropped).

p. 19: Drawing from John Claudius Loudon, *Encyclopaedia of Gardening* (Longman, Rees, Orme, Brown & Green, 1824), 1035, public domain (altered).

Garden of Relation: Drawing the Climatic Intelligence of Plants
p. 23, 26-27, 28-29: Drawings by Bonnie-Kate Walker used with permission.

Plants on the Move
p. 32: Image by Ryan Franchak, used with permission.

p. 36: "*Umbrella Tree (Magnolia tripetala)*" (1932) by Mary Vaux Walcott, Smithsonian American Museum of Art, Gift of the artist, 1970.355.57, public domain.

Green Gold: The Akkoub's Settler Ecologies
p. 39: "*Asteraceae: Gundelia tournefortii* (akkoub, tumble thistle)" by Cataloging Nature used under CC BY-2.0 license via Flickr.com (altered).

p. 43: Photograph by Irus Braverman used with permission, and "Spiny Gundelia" by Lehava Activity 2013 Pikiwiki Israel used under CC BY-2.5 license via Wikimedia Commons.

The Vault is a Bunker, The Arsenal Are Seeds
p. 44: Photograph by The Crop Trust used under CC BY-NC-SA 2.0 license (cropped).

p. 46, 48: Photographs by Xan Sarah Chacko used with permission, cropped.

p. 47: Photograph by Nuclear Regulatory Commission (2014) used under CC BY-2.0 Deed license, cropped.

p. 49: Image by Snøhetta used with permission (cropped).

p. 50: Photograph by Navdanya International used with permission (cropped).

p. 51: Photograph by Karen M'Closkey used with permission.

Design(ed) Decay
p. 52-53: Photograph by Gabriel Li used with permission.

p. 54, 56, 58: Photographs by Ally Schmaling used with permission.

In Conversation with Giovanni Aloi
p. 60: Photograph provided by Giovanni Aloi used with permission.

p. 62: "Rootbound #3, Exercises in Rootsystem Domestication" (2018) by Diana Scherer used with permission.

p. 65: "Bioremediation (Kudzu)" (2018–2019) by Jenny Kendler used with permission.

Forensic Ecologies and the Botanical City
p. 68-69: "Jardins de béton" (2012) by Groume used under CC BY-SA 2.0 license (cropped).

p. 72: Illustration by Richard Deakin, *Flora of the Colosseum of Rome* (1855), public domain.

Botanic Lessons from the Prairie
p. 74-75: Image by Jono Gilbert, used with permission (cropped).

In Conversation with Jared Farmer
p. 78: Image by Xinyu Liu using photograph by Jared Farmer (altered), used with permission.

p. 82: "The Sacred Bo-tree of Ceylon" by James Ricalton, public domain via *Scribner's Magazine* 10, no. 3 (1891): 321, and photograph by Gareth James (2016) used under CC BY-SA 2.0 license.

p. 83: Photograph courtesy of Ericson Collection, Cal Poly Humboldt University Library; Photograph by the National Park Service, public domain.

p. 84: Photograph by Ragesoss at the United States National Arboretum used under CC BY-SA 3.0 license (cropped), and photograph by Jared Farmer used with permission (cropped).

In Conversation with Patrick Blanc + Catherine Mosbach
p. 86: Image by Ryan Franchak used with permission.

p. 89: Drawing by Mosbach Paysagiste used with permission.

p. 91: Images courtesy of Patrick Blanc.

p. 92: Photograph by Mosbach Paysagiste used with permission.

Conceits and Constructs: Vegetal Architecture
p. 94: Image of Louis Sullivan entrance detail on the Carson Pirie Scott Building, Chicago, Historic American Buildings Survey, public domain via Wikimedia Commons.

p. 96: Image courtesy of Ruy-Klein used with permission, and image courtesy of Eco-Logic/ ©NAARO used with permission.

p. 101: Photographs of glass detail and greenhouse by Annette Fierro used with permission, and image of Ricola building courtesy of Margherita Spiluttini/ Architekturzentrum Wien, Collection used with permission.

p. 102: Image courtesy of AJN used with permission, and image of Musée de Quai Branly courtesy of Patrick Blanc used with permission.

Plant Samples

p. 104-105: Image generated using Midjourney by Maura McDaniel (2023) used with permission.

p. 108-109: NASA Earth Observatory images by Jesse Allen using data provided by the University of Maryland's Global Land Cover Facility and the U.S. Geological Surveys, public domain.

p. 110: Images by Maura McDaniel containing modified Copernicus Sentinel data (2023) processed by Sentinel Hub, used with permission.

p. 112-13: Images by Audrey Genest using LUMION (2023), used with permission.

Smart Plants and the Challenges of Multispecies Narrative

p. 115: "The Day of the Triffids" (1962) poster art, attributed to Joseph Smith for Allied Artists, entered the public domain 28 years after its US publication date due to failure to assert or renew copyright (altered).

IN THE NEXT ISSUE OF LA+

Issue 20 of LA+ Journal brings you the results of our fifth international design ideas competition. **LA+ EXOTIQUE** asked entrants to redesign the forecourt of the Museum of Natural History in Paris. The Museum–founded in 1793– sits within the Jardin des Plantes grounds, which include themed gardens, a zoo, and five themed galleries. In addition to its collections, the Museum is an active research institution studying the evolution of life on this planet. **LA+ EXOTIQUE** will showcase the award-winning designs and a comprehensive Salon des Refusés. The issue will also feature an essay by LA+ creative director Catherine Seavitt and interviews with jurors Julia Czerniak, Sonja Dümpelmann, Catherine Mosbach, Signe Nielsen, and Marcel Wilson.